It is widely believed that reproductive cycles are very similar between human females. However, there are in fact considerable variations both between individuals and within the reproductive life of any given individual. 'Normal' reproductive cycles cover a wide range of eventualities, and the likelihood of successful monthly egg release and ensuing pregnancy can be modified by a large number of factors.

In this book, the variability of human fertility is examined by first looking at the physiological processes regulating reproduction, and the roles of metabolic adaptation and metabolic load. Inter-population variation in normal ovarian function, is then discussed, covering the importance of factors such as age, disease, and breastfeeding in modifying ovarian function. This will be an important book for all those interested in human fertility.

D1638292

Cambridge Studies in Biological Anthropology 19

Variability in human fertility

Cambridge Studies in Biological Anthropology

Series Editors

G.W. Lasker
Department of Anatomy and Cell Biology,
Wayne State University,
Detroit, Michigan, USA

C.G.N. Mascie-Taylor
Department of Biological Anthropology,
University of Cambridge

D.F. Roberts
Department of Human Genetics,
University of Newcastle-upon-Tyne

R.A. Foley
Department of Biological Anthropology,
University of Cambridge

Variability in human fertility

EDITED BY

LYLIANE ROSETTA
(CNRS, France)

AND

C.G.N. MASCIE-TAYLOR
(University of Cambridge, UK)

CAMBRIDGE
UNIVERSITY PRESS

CAMBRIDGE UNIVERSITY PRESS
Cambridge, New York, Melbourne, Madrid, Cape Town, Singapore, São Paulo, Delhi

Cambridge University Press
The Edinburgh Building, Cambridge CB2 8RU, UK

Published in the United States of America by Cambridge University Press, New York

www.cambridge.org
Information on this title: www.cambridge.org/9780521117944

First published 1996
This digitally printed version 2009

A catalogue record for this publication is available from the British Library

ISBN 978-0-521-49569-1 hardback
ISBN 978-0-521-11794-4 paperback

Contents

viii   *Contents*

Contributors

Gillian R. Bentley
Department of Biological Anthropology, University of Cambridge,
Downing Street, Cambridge CB2 3DZ, UK

Peter T. Ellison
Department of Anthropology, Harvard University, Cambridge,
MA 02138, USA

M. L. Goubillon,
URA CNRS 1454, Université Lyon I, Lyon, France

Geoffrey A. Harrison
Institute of Biological Anthropology, University of Oxford,
58 Banbury Road, Oxford OX2 6QS, UK

P. C. Lee
Department of Biological Anthropology, University of Cambridge,
Downing Street, Cambridge CB2 3DZ, UK

Peter G. Lunn
Medical Research Council, Dunn Nutrition Laboratory, Milton Road,
Cambridge CB4 1XJ, UK

C. G. Nicholas Mascie-Taylor
Department of Biological Anthropology, University of Cambridge,
Downing Street, Cambridge CB2 3DZ, UK

Nicholas G. Norgan
Department of Human Sciences, Loughborough University,
Loughborough, Leicestershire LE11 3TU, UK

G. Plu-Bureau
Department of Reproductive Medicine, Hôpital Necker,
Paris, France

M. Rieu
Faculté de médecine Cochin-Port Royal,
24 rue du Faubourg Saint Jacques, 75014 Paris,
France

Lyliane Rosetta
Laboratoire de Physiologie des Adaptations,
Faculté de médecine Cochin-Port Royal,
24 rue du Faubourg Saint Jacques,
75014 Paris, France

Prakash S. Shetty
Human Nutrition Unit, Department of Public Health and Policy,
London School of Hygiene and Tropical Medicine, 2 Taviton Square,
London WC1E 7HT, UK

J-C Thalabard
URA CNRS 1454, Université Lyon I, Lyon, France

1 Introduction: the biological anthropological approach

G. A. HARRISON

Concern over world population growth has led to many investigations of human fertility including, on a global scale, the 'World Fertility Survey'. We certainly now know much more about the detailed nature of the 'population problem' than we did just a few years ago. But there are limitations to the broad-scale demographic methods which have been mainly used, especially in identifying the actual causes for fertility patterns, which are typically complex and interacting. Intense micro-studies are required for this, with detailed and prolonged social observations such as characterize anthropology.

Biological anthropology sees reproduction and fertility as part of an individual's whole life cycle and of a community's total ecology.

The ability to reproduce develops slowly over about a quarter of the life span, though much of the change is concentrated in a 3- or 4-year period. All stages are affected by various environmental influences, particularly nutrition, disease and altitude, as is well exemplified in the huge variation in the age of menarche and in its secular trend. During the mature years, fertility depends upon innumerable other factors than sexual behaviour; some of them obviously part of the reproductive process such as infant care practices, others acting indirectly through phenomena like maternal health, physical activity levels, and social life styles. It is important to take note of the fact that in many societies most adults, and particularly women, are working throughout the daylight hours and often into the night. Come old age, and specifically the menopause in women, reproductive roles can still exist, especially in providing all kinds of support for the care of offspring when they are reproducing. Both in providing supplementary child care and in relieving economic tasks, this support can be crucial to reproductive success. Throughout life then there are environmental determinants of an individual's fertility, but for most of the time most people are doing much more than reproducing!

This takes one inevitably into consideration of the ecology of communities, as it relates to fertility. Here one needs to consider a wider range of

1

biological variables such as the effects of seasonality on work regimes and nutritional health, and the endemicity of infectious disease on infant mortality and the possibilities for reproductive compensation. But, with an ecological focus, one also needs to take account of a whole spectrum of economic and social factors. These will determine the overall quality of the environment, the daily regimes and the pattern of the life cycle changes. They will also affect such highly influential factors of fertility as age of marriage, degrees of spouse separation and, perhaps, most importantly of all, attitudes to reproductive behaviour and family life.

Economy and social factors also play a significant role in determining how resources are allocated within a society and, indeed, within a family. All human societies are stratified to some degree, and some individuals and families achieve access to much more favourable environments than others. This is bound to have significance for reproductive success. However, whilst in other animals the best environment for survival will almost inevitably be the best one also for reproduction, this is not necessarily the case for humans. Added to all the environmental determinants of fertility, in humans there is also the matter of 'choice'. To varying degrees, and almost completely in many developed societies, sexual behaviour can be decoupled from having children. I am not aware of a full review in the anthropological literature of the contraceptive techniques that are used in different societies, but without doubt they are various and one or other probably universally practised. And, if abortificants and the like are not used, there is always infanticide or abstinence! The point, however, to be made here is that, in studies of fertility, it is important to try and determine what reproductive strategies people are attempting to practise and why. Clearly, the social dimension is critical to this.

It must, of course, not be presumed that all people, even in a simple society or stratum of one, have the same strategy. There will undoubtedly be variance in biological 'performance' but there will also be variance in personal 'objectives', and the latter will depend upon complex components of family structure and wealth, both past and present.

A general focus on within population variance is extremely important. Methodologically it provides the opportunity for analysing causation, but it is also the raw material for natural selection. Even when there is no genetic component to the variance, the action of natural selection is of significance and can change the form of reproductive strategies. Variability of sibship size is also of interest for assessing the role of stochastic processes in the demogenetics of populations; it is a key component of effective population size.

These issues indicate why biological anthropologists are so concerned with 'within' population or 'within' society situations. In the past the focus tended to be on comparative population studies and the nature and causes for 'between' population differences. This is still so in some areas of demographic study, but ultimately it is what happens within groups that determines the differences between them. And it is only in the within situation that one can make rigorous deterministic analyses.

However, as I have already intimated, studies of this kind are demanding of time, money and commitment. They involve long periods of field study and careful longitudinal observation. They demand good knowledge of local languages and customs and an ability of the investigator to become at least partly integrated into the society being studied. Formal pre-devised questionnaires may be nearly useless, and time and patience is required to conduct in-depth interviews, not only of reproductive histories but also of household economy and ecology, customs and belief systems and individual aspirations. Even establishing the age of subjects can be a difficult and tedious business, and diary approaches, even if they can be used at all, are going to give very unrepresentative information. Often, it will be extremely helpful to undertake physiological and endocrinological monitoring, especially of the hormones controlling fecundity, and such monitoring needs to be done to take account of seasonal and other systematic ecological changes.

A formidable task indeed, but one that has begun to be undertaken in a number of situations (as this book demonstrates well). They already show the value of the anthropological approach, but they also highlight a critical methodological problem long known to field workers: the problem of sample size. Clearly, in the real world of investigation there has always to be a compromise between intensive investigations of a few subjects and extensive investigation of many. In some traditional societies, even total ascertainment studies are fraught with the problem of detecting generality. We need to address these problems more carefully in biological anthropology, particularly by identifying mixed plan strategies in which the more difficult time allocation and physiological observations are based on a few, and the basic demographic data collected on as many as possible. The latter in many cases will provide the knowledge for identifying those who most meaningfully can be intensively studied. But more thought should be given to this, and in advance, rather than in the field!

Part I:
Hormonal aspects of fertility regulation

2 The hypothalamo-pituitary regulation of the reproductive function: towards an increasing complexity

J. C. THALABARD, M. L. GOUBILLON
AND G. PLU-BUREAU

For the historian of sciences, the state of knowledge on a specific biological topic may be viewed as a complex process characterized by an alternance of phases of accumulation of experiments and their corresponding results, respectively designed and discussed in a given paradigm, and phases when the scientific community is forced to revise the current paradigm and propose a new one more able to give a general framework for the interpretation of past and present data and design of future studies. The change to a new paradigm is frequently associated with the emergence of new tools and a change of scale in the understanding of the phenomena (Kuhn, 1970).

A similar analysis can be applied to our understanding of the reproductive axis. Schematically, the last two decades have been associated with the isolation of the three key structures involved in the reproductive axis in mammals, i.e. the gonads, the anterior pituitary and the hypothalamus and the identification of the roles, together with the description, of the urinary and plasma dynamic patterns of their principal secretions, and associated metabolites. The current period is characterized by both an internal and external complexification of the system. The internal complexification corresponds to a better understanding of the numerous intermediate steps involved in the response to an hormonal stimulus eventually resulting in another secretion by the stimulated cell. The external complexification corresponds to: (i) the importance of the hypothalamo-pituitary target as a key structure: (ii) the demonstration of the effects on the reproductive axis of factors normally associated with other physiological functions; (iii) the discovery of new factors modulating, simultaneously, two or more different physiological functions; (iv) the necessity of giving up the simplistic view of

7

one neuron-one neuro-transmitter, considering the importance of the colocalization phenomenon; (v) the emergence of non-receptor mediated phenomenon; (vi) the importance of the temporal sequence of neurotransmitter activations and the different time-scales among the differing interplaying actors.

The aim of this limited presentation is to review, in an non-exhaustive manner, some recent perspectives on the hypothalamo-pituitary-gonadal axis (HPG) in mammals, in order to emphasize this current phasis of non-hierarchized complex state of scientific knowledge. Finally, implications for the effects of food and energy consumption will be briefly discussed.

The GnRH pulse generator

Extensive studies on the molecular and cellular aspects associated with GnRH have been achieved: from the evolutionary point of view, the number of isolated forms of GnRH among species remains very limited, the human form of GnRH having appeared, probably, around 400 million years ago. This fact suggests the importance of this peptide for the reproductive function, and hence survival of the species (Sherwood, Lovejoy & Coe, 1993), although GnRH releasing neurons do not appear to control only the HPG. The GnRH neurons originate from the olfactory placode outside of the brain during early development and migrate into the forebrain. From early neurogenesis, GnRH neurons seem to be already committed (El Amraoui & Dubois, 1993). Only 50–70% of the axons of the GnRH neurons terminate on the median eminence of the hypothalamus, and therefore can release GnRH into the portal vessels. GnRH exists as part of a larger precursor protein composed of a signal peptide followed by a GnRH associated peptide (GAP). GAP has been shown to coexist with GnRH in hypothalamic neurons and may act as a conformational assistant, without any other known physiological role. Other locations of GnRH production outside the central nervous system have been demonstrated, such as in the follicular fluid or in the testis, but, except for a local autocrine/paracrine role, their physiological contribution remains unclear. Immunocytochemical studies have shown synaptic connections between GnRH neurons, which are therefore forming a specific network.

The temporal pattern of release of the gonadotropin hormone-releasing hormone (GnRH) into the portal vessels by a subgroup of hypothalamic neurons represents the final common signal for the control of the HPG (for review see Knobil, 1989). In humans, the possibility of characterizing the

GnRH intermittent release into the portal vessels can be achieved only by means of the determination of the plasma luteinizing hormone (LH) levels through intensive blood sampling over extended period of time. New and more sensitive radioimmunometric methods for plasma LH measurement have forced the simplistic view of a quiescent pulse generator structure during childhood to be reconsidered, as it turns out that pulses of LH can be detected not only at birth and after puberty but at almost any age during childhood, the critical factor becoming the quality or efficiency of the releasing pattern (Apter *et al.*, 1993).

Animal studies provide, in addition to intensive blood sampling, other and more direct accesses to the GnRH releasing system. The multiunit activity (MUA) electrical recording of the arcuate nucleus neurons in monkeys (Wilson *et al.*, 1984), and, more recently in goats (Ito, Tanaka & Mori, 1993) or rats (Kimura *et al.*, 1991; Goubillon *et al.*, 1995), provides a real-time follow-up of this system. The continuous recording of this signal during the menstrual cycle in female monkeys has shown an unexpected and dramatic reduction of the GnRH pulse generator activity during the pre-ovulatory LH surge, which is associated with the sharp rise in the plasma oestradiol level (O'Byrne *et al.*, 1991). Indeed, administration of oestradiol either in intact or ovariectomized female monkeys or ovariec-tomized rats induces a sharp decline in GnRH associated MUA, at least partly mediated by endogenous opiates (Grosser *et al.*, 1993; Kato, Hiruma & Kimura, 1994). The possibility of collecting intensive portal blood samples provides another approach to the system, as shown in sheep by Caraty & Locatelli (1988), Moenter *et al.* (1991), Clarke *et al.* (1993). In parallel to these *in vivo* studies, *in vitro* dynamical studies tried to localize and characterize the isolated hypothalamic pulse generator, using either hypothalamic explants (Bourguignon *et al.*, 1993; Rasmussen, 1993) or immortalized GnRH cell lines (Mellon *et al.*, 1990; Hales, Sanderson & Charles, 1994). One of the main outcome of these studies is the hypothesis that the intermittent GnRH release by the GnRH neuronal network is the resultant of an intrinsic phenomenon of the GnRH neuron.

However, the activity of this 'pace-maker', generating the release of this intermittent temporal pattern of GnRH, is exposed to a large number of modulators. Morphological studies have shown that hypothalamic neur-ons are usually not restricted to only one neurotransmitter but colocalize several other chemical messengers. They receive a significant number of afferent fibers from other neuronal populations. In addition, a rich fiber system surrounds their terminals which may interfere either with the release of GnRH into the portal blood (Halasz, 1993), or with its action at the pituitary level, by the release of neurohormones into the portal vessels.

The hypothalamic regulation of the GnRH system

In addition to the modulation exerted by endogenous opiates (Petraglia, Vale & Rivier, 1986; Williams *et al.*, 1990*a*; Conover *et al.*, 1993) and catecholamines, other neurotransmitters have emerged as important regulating factors.

Amino acid neurotransmitters

The action of the family of the amino acid neurotransmitters (AANT) (glutamate, aspartate, homocysteic acid, gaba-aminobutyric acid -GABA-, glycine and taurine) on the reproductive axis currently is studied intensively.

Glutamate and aspartate

These amino acids belong to the subfamily of the excitatory amino acids (EAAs), the effects of which involve three different types of receptors. EAA- secreting neurons together with EAA receptors are found in the hypothalamus and other brain area (Meeker, Greenwood & Hayward, 1994). In monkeys, Goldsmith *et al.* (1994) have shown that the GnRH network receives synaptic inputs from glutamate-immunoreactive neurons. *In vitro*, glutamate stimulates GnRH release in immortalized GT-1 GnRH neurons (Spergel *et al.*, 1994). *In vivo*, the net effect of glutamate or aspartate, given either centrally or peripherally, is a rapid stimulation of the GnRH release, which depends on the steroidal milieu (Arias *et al.*, 1993; Herbison, Robinson & Skinner, 1993). Repetitive injections elicit a pattern of pulsatile LH secretion in rats and primates, able to stimulate the gonadal axis in pre-pubertal animals (Plant *et al.*, 1989; Nyberg *et al.*, 1993). Their physiological role seems to be related to the process of sexual maturation. It has been suggested also a role in the control of the GnRH release at the time of ovulation (Ping *et al.*, 1994) and in the control of reproduction in seasonal breeder species. In addition, this stimulation is markedly decreased in lactating rats exposed to an intense suckling stimulus (Pohl, Lee & Smith, 1989; Abbud & Smith, 1993). It is assumed to play a critical role in the expression of pre-programmed changes in neuroendocrine functions, which occur over an individual's life-span (Cowell, 1993).

GABA

GABA is generally considered as having an inhibitory action. However, recent *in vitro* studies on immortalized hypothalamic GnRH neurons suggest a biphasic action characterized by an initial sharp GnRH release, mediated by the GABA-A receptor (Hales, Sanderson & Charles, 1994) followed by a sustained inhibition, mediated by the GABA-B receptor (Martinez de la Escalera, Choi & Weiner, 1994). *In vivo* studies suggest a role of GABA in mediating the adrenergic and EAA-induced increase in GnRH release in rats (Akema & Kimura, 1993) and the changes in the oestrogen negative feedback effects on the hypothalamus in ewes (Scott & Clarke, 1993).

Galanin (GAL)

Galanin is a widely distributed neuropeptide, highly conserved through the evolution. A subgroup of Gal neurons projects axons to the external zone of the median eminence and colocalizes GnRH (Merchenthaler, Lopez & Negro-Vilar, 1990; Lopez *et al.*, 1991). An oestrogenic milieu increases its coexpression, while progesterone decreases it. However, after prolonged exposure to oestradiol in ovariectomized rats, progesterone further increases the GAL mRNA level in correlation with the gonadotropin-induced surge (Brann, Chorich & Mahash, 1993). Other subgroups of GAL neurons, bearing oestrogen receptors, colocalize noradrenalin or dopamine and could be involved in the aging of the reproductive system (Ceresini *et al.*, 1994). In animal studies, GAL seems to act at the pituitary level as it stimulates GH and Prolactin secretion (Wynick *et al.*, 1993), potentiates GnRH release and inhibits the stress-induced ACTH release. Studies in humans remain limited and have only shown a stimulation of GH secretion, without any effect on LH and FSH (Carey *et al.*, 1993).

NPY

Neuropeptide Y has emerged as a physiologically significant contributor to the control of LH secretion (Freeman, 1993). It is a member of the pancreatic peptides, which are now found in the central nervous system. One of its interests is its involvement in the control of food intake (Li, Hisaon & Daikoku, 1993; Wilding *et al.*, 1993). Its structure presents a large degree of species homology. The corresponding NPY-neurons receive adrenergic afferences and part of them project their axons into relevant hypothalamic sites. However, the existence of direct contacts between GnRH- and NPY-neurons seems species-dependent (Thind, Boggan & Goldsmith, 1993). The activity of NPY-neurons depends on the steroid milieu. A possible role of NPY in potentiating the GnRH release at the time of the onset of puberty has been shown in female rhesus monkeys' (Gore, Mitsushima & Terasawa, 1993). In the decreased caloric intake lamb model, which delays the onset of puberty, Prasad *et al.* (1993) did not find a change in NPY release from the median eminence, whereas they observed both a decrease in LHRH and β-endorphin release, as compared with control animals. The use of NPY-antiserum in food-deprived rats induces an increase in both amplitude and frequency of growth hormone (GH) suggesting its involvement in mediating the GH-inhibition in food deprivation (Okada *et al.*, 1993).

Cytokines

Various chronic inflammatory and stress disorders are known to be frequently associated with alteration of the reproductive function. These states mobilize components of the immune system, like interleukines (IL). The production of IL and their corresponding receptors is modulated by glucocorticoids (Betancu *et al.*, 1994). IL-1β decreases the GnRH release and synthesis (Rivest *et al.*, 1993a). This effect seems partly mediated by the stimulation by interleukines of the prostaglandin-, corticotropin-releasing hormone- (CRF) and vasopressin- (VP) pathways (Rivier & Erickson, 1993; Shalts *et al.*, 1994). The inhibitory action of CRF on the GnRH release has been shown to be mediated, as least partly, by endogenous opiates (Olster & Ferin, 1987; Williams *et al.*, 1990b, Rivest, Plotsky & Rivier, 1993b). Bacterial endotoxin can directly activate the release of IL, which activates the HPA axis and inhibits the HPG axis. Thymocytes, which are part of another component of the immune system, have been shown to produce a peptidic factor, which stimulates the hypothalamic GnRH release; the physiological importance of this phenomenon remains unknown.

This non-exhaustive list of neuropeptides represents the components of a complex spatio-temporal patterning of peptidergic signaling orchestrated by the gonadal steroids and other gonadal secretions. Other phenomena, like the neuronal plasticity and the concept of non-receptor mediated regulation are additional elements of this system. In this regard, nitric oxide (NO) emerges as a new class of diffusible neurotransmitter implied in the control of the synaptic communication. The efflux of NO is stimulated by activation of NMDA-receptors and administration of inhibitor of the NO synthase markedly suppresses the progesterone- induced surge in oestrogen-primed ovariectomized rats; in rat hypothalami *in vitro*, spontaneous release of NO increases GnRH release, whereas administration of NO inhibitor alters the NMDA-induced GnRH release (Bonavera *et al.*, 1993). Its importance in the regulation of the HPG remains to be elucidated.

The pituitary–gonadal modulation of the GnRH system

In adult, a necessary condition for a normal operation of the pituitary–gonadal system is the presence of a hypothalamic regular GnRH 'pacemaker'. However, experimental studies have shown that the peripheral secretions of the HPG do affect the expression and regularity of the hypothalamic pulse generator.

The pituitary

The pituitary can be viewed as a complex filter–amplifier system, composed of different types of intermingled cells, with specialized secretions in response to the conjunction of neurohormonal and peripheral stimuli. The major determinants of the HPG at this level remain the GnRH stimulus and the sexual steroids. To date, few evidences of short-loop feedback of pituitary secretions on the GnRH system have been shown (Selvais *et al.*, 1993). However, some of the substances, previously mentioned, add to their action at the hypothalamic level an enhancing effect at the pituitary level, such as NPY, which has a potentiating effect on the GnRH stimulus, dependent on the steroid environment (O'Conner *et al.*, 1993; Bauer-Dantoin *et al.*, 1993*a,b*) or vasoactive intestinal peptide (VIP), which increases specifically the LH response to GnRH (Hammond *et al.*, 1993). Bacterial endotoxin appears to stimulate the adrenal axis and decrease the LH response in sheep (Coleman *et al.*, 1993). In addition to genetic factors, the modalities of the stimulation of the gonadotropin cells may affect the molecular forms of the secreted hormones (Stanton *et al.*, 1993; Wide & Bakkos, 1993), which may affect the reproductive function (Weiss *et al.*, 1992; Papandreou *et al.*, 1993).

The gonads

From the endocrine point of view, the ovaries present the remarkable feature of a limited functional life span in contrast to other endocrine organs. Their secretions depend on the availability of a non-renewable stock of follicles, although the oestrogen secretion does not seem to be severely modified until late in the reproductive life. In humans, at birth, the number of healthy non-growing follicles (NGF) is approximately 130 000–500 000 per ovary and is limited to a very small residue when the women reaches her menopause. The decay rate in NGF, which starts at birth, has been shown to be accelerated several years before the menopause, at around 35 years (Gougeon, Ecochard & Thalabard, 1994). A rise in mean

FSH level has been reported at approximately the same age, suggesting an early adaptation of the HPG to the modifications of the ovarian state.

It should be noted that the determination of the structure of the LH/ FSH receptor family has initiated new approaches in the understanding of the regulation of the action of the gonadotropin hormones on their target cells (Segaloff & Ascoli, 1993).

Sexual steroids

Oestradiol and progesterone remain the principal ovarian regulators of the HPG. Oestradiol exerts its action locally at the ovarian level but also at the pituitary and the central nervous system levels. Different modalities of action of oestrogens on their target cells on different time-scales ranging from a few seconds to several hours or weeks have been described. Short-term biochemical events triggered by oestrogens are suggestive of direct membrane actions (Morley *et al.*, 1992), where some long-term effects might be the consequences of morphological remodelling affecting both glial cells and neuronal connectivity (Naftolin *et al.*, 1993; Shughue & Dorsa, 1993). The absence of oestrogen receptors on the hypothalamic GnRH neurons prevents theoretically a direct action of oestrogen on the GnRH release, but oestradiol receptors have been localized in various sites of the brain, including the hypothalamic area, in both neurons and glial cells. The loss of the oestrogen milieu progressively affects the sensitivity of the HPG (Rossmanith, Reichelt & Scherbaum, 1994). Endogenous opiates have been shown to mediate some of the actions of oestradiol and progesterone on the hypothalamus.

Inhibin and related peptides

Major progress in the past few years has been achieved, regarding the biophysical and biochemical properties of inhibin and related peptides, which compose a new family of active compounds. Inhibin is a 58 kD dimeric glycoprotein composed of two subunits α and β, the latter corresponding to two distinct types A and B. There is a potentially large variety of circulating inhibin species, with a preferential secretion of inhibin A or inhibin B under various circumstances. This particularity explains the difficulty in measuring the circulating inhibin. The related peptides are: first, activin (dimer β–β), which stimulates FSH activities; second, follistatin or FSH suppressing activity, which is structurally unrelated to inhibin but has a similar action and; third, the inhibin binding proteins, which are important to know as they may interfere with inhibin measurement.

Bioassays for inhibin using the property of inhibiting FSH activity have been developed. More recently, heterologous or more specific (αC subunit) RIAs have been developed. The reference is so far a porcine inhibin standard (Melbourne); but the assay sensitivity remains poor (60–70 mU/ml). The clinical applications remain limited. In adult females, inhibin appears to be a FSH- regulating hormone. The related peptides seem to tightly regulate the FSH release by pituitary cells to incoming stimuli (Bilezikjian *et al.*, 1993*a,b*). It starts rising at the end of the follicular phase, and shows an even greater peak during the luteal phase, closely related with the rise and fall of progesterone. In ageing women, the inhibin response to ovarian stimulation falls whereas the oestradiol response persists (Burger, 1993).

IGF-1

Insulin-growth factor-1 is an ubiquitous growth-factor, with a peripheral insulin-like action that has been shown to have paracrine actions at different levels of the HPG axis. At the ovarian level, it stimulates the growth of human granulosa cells, and the secretion of oestradiol, even in the absence of FSH, and amplifies the effect of FSH (Mason *et al.*, 1993). At the hypothalamic level, it may interact with the N-methyl-D-aspartate receptors, thus antagonizing the stimulation of GnRH release (Bourguignon *et al.*, 1993). IGF-1 bioactivity is regulated by specific binding proteins (IGFBP), with different affinities for IGF-1. IGFBP-1 and IGF-1 have been hypothetized as peripheral signals reflecting the metabolic status and available caloric reserves at the hypothalamic and ovarian levels (Jenkins *et al.*, 1993).

Effect of food intake and exercise

Severe chronic energy restriction as observed in case of a serious increase in energy expenditure and/ or major restriction in caloric intake are known to be associated with alteration of the HPG (Schweiger *et al.*, 1992). Further studies on the neuroendocrine effects of acute or chronic exercise in eumenorrheic or amenorrheic women seem to confirm a modification of the endogenous opioid tone (Walberg, Franke & Gwazdauskas, 1992) and/ or the adrenal axis (Kanaley *et al.*, 1992). In contrast, the mechanisms involved in food restriction appear somehow different. A series of experimental studies in male monkeys have shown that: (i) a short-term restriction in daily meal intake is rapidly followed by a slowing down of the

GnRH pulse generator activity and a decrease of testosterone secretion (Cameron *et al.*, 1991); (ii) the restoration of which depends on the size and timing of the refeed meal (Parfitt, Church & Cameron, 1991; Mattern *et al.*, 1993); (iii) the effect seems directly related to the absence of a nutritional/metabolic signal and not to the psychological stress induced by food withdrawal (Schreihofer, Golden & Cameron, 1994*a*; Schreihofer, Amico & Cameron 1993*b*); (iv) neither the adrenal axis and endogenous opiates nor intestinal peptide cholecystokinin appears to be involved in mediating this effect (Helmreich & Cameron, 1992; Helmreich, Mattern & Cameron, 1993; Schreihofer, Parfitt & Cameron, 1993*c*).

Other metabolic signals associated with the energy balance and metabolic 'stress' (Mikines *et al.*, 1989) have been shown to alter the GnRH signal, such as insulin-induced hypoglycemia in monkeys (Chen *et al.*, 1992; Heisler *et al.*, 1993), the effect of which depends on the presence of the ovaries. Further studies are needed to confirm that these short-term alterations of the GnRH releasing pattern may durably and visibly affect the HPG.

In addition, food composition may be the source of metabolites interfering with the normal operation of the reproductive axis. For instance, high fibre diets are associated with a production of molecules by intestinal bacteria (lignans) close to the oestrogen structure and may facilitate regular cycles, as suggested by Phipps *et al.* (1993) in a randomized trial comparing flax seed supplemented- versus not-supplemented- normal healthy women.

Conclusion

The direct interest of the biological anthropologist is the human being in his natural environment, and there remains a large gap between experimental data on laboratory models of stress or stimulation of immune system and the complexity of the factors interfering in the interpretation of human observations in field studies. Current advances in the understanding of the reproductive axis do not add, presently, new methods of quantification for the field worker but, simply, provide experimental evidences for biological connections and interrelationships between systems, which the human anthropologist, based on his expertise, was already used to consider as interconnected.

REFERENCES

Abbud, R. & Smith, M. S. (1993). Altered luteinizing hormone and prolactin responses to excitatory amino acids during lactation. *Neuroendocrinology*, **58**, 454–64.

Akema, T. & Kimura, F. (1993). Differential effects of GABAA and GABAB receptor agonists on NMDA-induced and noradrenaline-induced luteinizing hormone release in the ovariectomozed oestrogen-primed rat. *Neuroendocrinology*, **57**, 28–33.

Apter, A., Bützow, T. L., Laughlin, G. A. & Yen, S. S. C. (1993). Gonadotropin-releasing hormone pulse generator activity during pubertal transition in girls: pulsatile and diurnal patterns of circulating gonadotropins. *Journal of Clinical Endocrinology and Metabolism*, **76**, 940–9.

Arias, P., Jarry, H., Leonhardt, S., Moguilevsky, J. A. & Wuttke, W. (1993). Estradiol modulates the LH release response to *N*-methyl-D-aspartate in adult female rats: studies on hypothalamic luteinizing hormone-releasing hormone and neurotransmitter release. *Neuroendocrinology*, **57**, 710–15.

Bauer-Dantoin, A. C., Knox, K. L., Schwartz, N. B. & Levine, J. E. (1993a). Estrous cycle stage-dependent effects of neuropeptide-Y on luteinizing hormone (LH)-releasing hormone-stimulated LH and follicle-stimulating hormone secretion from anterior pituitary fragments *in vitro*. *Endocrinology*, **133**, 2413–17.

Bauer-Dantoin, A. C., Tabesh, B., Norgle, J. R. & Levine, J. E. (1993b). RU486 administration blocks neuropeptide Y potentiation of luteinizing hormone (LH)-releasing hormone-induced surges in proestrous rats. *Endocrinology*, **133**, 2418–23.

Betancu, C., Lledo, A., Borrell, J. & Guazza, C. (1994). Corticosteroid regulation of IL-1 Receptors in the mouse hippocampus: effect of glucocorticoid treatment, stress, and adrenalectomy. *Neuroendocrinology*, **59**, 120–8.

Bilezikjian, L. M., Vaughan, J. M. & Vale, W. M. (1993a). Characterization and the regulation of inhibin/activin subunit proteins of cultured rat anterior pituitary cells. *Endocrinology*, **133**, 2545–53.

Bilezikjian, L. M., Corrigan, A. Z., Vaughan, J. M. & Vale, W. M. (1993b). Activin-A regulates follistatin secretion from cultured rat anterior pituitary cells. *Endocrinology*, **133**, 2554–60.

Bonavera, J. J., Sahu, A., Kalra, P. S. & Kalra, S. P. (1993). Evidence that nitric oxide may mediate the ovarian steroid-induced luteinizing hormone surge: involvement of excitatory amino acids. *Endocrinology*, **133**, 2481–7.

Bourguignon, J. P., Gerard, A., Gonzalez, M. L. A. & Franchimont, P. (1993). Acute suppression of gonadotropin hormone-releasing hormone secretion by insulin-like growth factor I and subproducts: an age-dependent endocrine effect. *Neuroendocrinology*, **58**, 525–30.

Brann, D. W., Chorich, L. P. & Mahesh, V. B. (1993). Effect of progesterone on galanin mRNA levels in the hypothalamus and the pituitary: correlation with the gonadotropin surge. *Neuroendocrinology*, **58**, 531–8.

Burger, H. G. (1993). Clinical utility of inhibin measurements. *Journal of Clinical Endocrinology and Metabolism*, **76**, 1391–8.

Cameron, J. L. & Nosbisch, C. (1991). Suppression of pulsatile luteinizing hormone and testosterone secretion during short term restriction in the adult male rhesus monkey (*Macaca mulatta*). *Endocrinology*, **128**, 1532–40.

Cameron, J. L., Weltzin, T. E., McConaha, C., Helmreich, D. L. & Kaye, W. H. (1991). Slowing of pulsatile luteinizing hormone secretion in men after forty-eight hours of fasting. *Journal of Clinical Endocrinology and Metabolism*, **73**, 35–41.

Caraty, A. & Locatelli, A. (1988). Effect of time after castration on secretion of LHRH and LH in the ram. *Journal of Reproduction and Fertility*, **82**, 263–9.

Carey, D. G., Ismaa, T. P., Ho, K. Y., Rajkovic, I. A., Kelly, J., Kraegen, Ferguson, AS Inglis, Shine, J. & Chisholm, D. J. (1993). Potent effects of human galanin in man: growth hormone secretion and vagal blockade. *Journal of Clinical Endocrinology and Metabolism*, **77**, 90–3.

Ceresini, G., Merchenthaler, A., Negro-Vilar, A. & Merchenthaler, I. (1994). Aging impairs Galanin expression in luteinizing hormone-releasing hormone neurons: effect of ovariectomy and/or estradiol treatment. *Endocrinology*, **134**, 324–30.

Chen, M. D., O'Byrne, K. T., Chiappini, S. E., Hotchkiss, J. & Knobil, E. (1992). Hypoglycemic 'stress' and gonadotropin-releasing hormone pulse generator activity in the rhesus monkey: role of the ovary. *Neuroendocrinology*, **52**, 133–7.

Clarke, I., Jessop, D., Millar, R., Morris, M., Bloom, S., Lightman, S., Coen, C. W., Lew, R. & Smith, I. (1993). Many peptides that are present in the external zone of the median eminence are not secreted into the hypophysial portal blood of sheep. *Neuroendocrinology*, **57**, 765–75.

Coleman, E. S., Elsasser, T. H., Kemmpainen, R. J., Coleman, D. A. & Sartin, J. L. (1993). Effect of endotoxin on pituitary hormone secretion in sheep. *Neuroendocrinology*, **58**, 111–22.

Conover, D. C., Kuljis, R. O., Rabii, J. & Advis, J. P. (1993). Beta-endorphin regulation of luteinizing hormone-releasing hormone release at the median eminence in ewes: immunocytochemical and physiological evidence. *Neuroendocrinology*, **57**, 1182–95.

Cowell, A. M. (1993). Excitatory amino acids and hypothalamo-pituitary–gonadal function. *Journal of Endocrinology*, **139**, 177–82.

El Amraoui, A. & Dubois, P. M. (1993). Experimental evidence for an early commitment of gonadotropin-releasing hormone neurons, with special regard to their origin from the ectoderm of nasal cavity presumptive territory. *Neuroendocrinology*, **57**, 991–1002.

Falsetti, L., Pasinetti, E., Mazzani, M. D. & Gastaldi, A. (1992). Weight loss and menstrual cycle: clinical and endocrinological evaluation. *Gynecology and Endocrinology*, **6**, 49–56.

Freeman, M. E. (1993). Neuropeptide Y: a unique member of the constellation of Gonadotropin- releasing hormones. *Endocrinology*, **133**, 2411–12.

Goldsmith, P. C., Thind, K. K., Perera, A. D. & Plant, T. M. (1994). Glutamate-immunoreactive neurons and their gonadotropin-releasing hormone–neuronal interactions in the monkey hypothalamus. *Endocrinology*, **13**, 858–68.

Gore, A. C., Mitsushima, D. & Terasawa, E. (1993). A possible role of neuropeptide Y in the control of the onset of puberty in famel rhesus monkeys. *Neuroendocrinology*, **58**, 23–34.

Goubillon, M. L., Van Hoeke, M. J., Kaufman, J. M. & Thalabard, J. C. (1995). Hypothalamic multiunit activity luteinizing hormone release in the castrated male rat. *European Journal of Endocrinology*, **133**, 585–90.

Gougeon, A., Ecochard, R. & Thalabard, J. C. (1994). Age related changes of the population of human ovarian follicles: increase in the disappearance rate of non-growing and early-growing follicles in aging women. *Biology of Reproduction*, **50**, 653–63.

Grosser, P. M., O'Byrne, K. T., Williams, C. L., Thalabard, J. C., Hotchkiss, J. & Knobil, E. (1993). Effects of naloxone on oestrogen-induced changes in hypothalamic gonatropin-releasing hormone pulse generator activity in the rhesus monkey. *Neuroendocrinology*, **57**, 115–19.

Halasz, B. (1993). Neuroendocrinology in 1992. *Neuroendocrinology*, **57**, 1196–207.

Hales, T. G., Sanderson, M. J. & Charles, A. C. (1994). GABA hax excitatory actions on GnRH-secreting immortalized hypothalamic (GT 1–7) neurons. *Neuroendocrinology*, **59**, 297–308.

Hammond, P. J., Talbot, K., Chapman, R., Ghatei, M. A. & Bloom, S. R. (1993). Vasoactive intestinal peptide, but not pituitary adenylate cyclase-activating peptide, modulates the responsiveness of the gonadotroph to LHRH in man. *Journal of Endocrinology*, **137**, 529–32.

Heisler, L. E., Pallotta, C. M., Reid, R. L. & Van Vugt, D. A. (1993). Hypoglycemia-induced inhibition of luteinizing hormone secretion in the rhesus monkeys is not mediated by endogenous opioid peptides. *Journal of Clinical Endocrinology and Metabolism*, **76**, 1280–5.

Helmreich, D. L. & Cameron, J. L. (1992). Suppression of luteinizing hormone secretion during food restriction in male rhesus monkeys (*Macaca mulatta*): failure of naloxone to restore normal pulsatility. *Neuroendocrinology*, **56**, 464–73.

Helmreich, D. L., Mattern, L. G. & Cameron, J. L. (1993). Lack of a role of the hypothalamic–pituitary–adrenal axis in the fasting-induced suppression of luteinizing hormone secretion in adult male rhesus monkeys (*Macaca mulatta*). *Endocrinology*, **132**, 2427–37.

Herbsion, A. E., Robinson, J. E. & Skinner, D. C. (1993). Distribution of estrogen receptor-immunoreactive cells in the preoptic area of the ewe: co-localization with glutamic acid decarboxylase but not luteinizing hormone-releasing hormone. *Neuroendocrinology*, **57**, 751–9.

Ito, K., Tanaka, T. & Mori, Y. (1993). Opioid peptidergic control of gonadotropin-releasing hormone pulse generator activity in the ovariectomized goat. *Neuroendocrinology*, **57**, 634–9.

Jenkins, P. J., Ibanez-Santos, X., Holly, J., Cotterill, A., Perry, L., Wolman, R., Harries, M. & Grossman, A. (1993). IGFBP-1: a metabolic signal associated with exercise-induced amenorrhea. *Neuroendocrinology*, **67**, 600–4.

Kanaley, J. A., Boileau, R. A., Bahr, J. M., Mismer, J. E. & Nelson, R. A. (1992). Cortisol levels during prolonged exercise: the influence of menstrual phase and menstrual status. *International Journal of Sports*, **13**, 332–6.

Kato, A., Hiruma, H. & Kimura, K. (1994). Acute estradiol modulation of electrical activity of the LHRH pulse generator activity in the ovariectomized rat: restoration by naloxone. *Neuroendocrinology*, **59**, 426–31.

Kimura, F., Nishihara, M., Hiruma, H. & Funabashi, T. (1991). Naloxone increases the frequency of the electrical activity of luteinizing hormone releasing hormone pulse generator in long-term ovariectomized rats. *Neuroendocrinology*, **53**, 97.

Kimura, F., Sano, A., Hiruma, H. & Funabashi, T. (1993). Effects of gamma-aminobutyric acod-A receptor antagonist, bicuculline, on the electrical activity of luteinizing hormone-releasing hormone pulse generator in the ovariectomized rat. *Neuroendocrinology*, **57**, 605–14.

Knobil, E. (1989). In *Control of the Onset of Puberty III*, eds. Delamarre-van de Waal, H. A., Plant, T. M., van Rees, G. P. & Shoemaker, J. (Excerpta Medica, New York), pp. 11–20.

Kuhn, T. (1970). *The Structure of Scientific Revolutions.* 2nd Edition. French translation, 1983. Flammarion Ed. Paris.

Li, S., Hisaon, S. & Daikoku, S. (1993). Mutual synaptic associations between neurons containing neuropeptide Y and neurons containing enkephalin in the arcuate nucleus of the rat hypothalamus. *Neuroendocrinology*, **57**, 306–13.

Lopez, F. J., Merchenthaler, I., Ching, M., Wiesniwski, M. G. & Negro-Vilar, A. (1991). Galanin: a hypothalamic-hypophysiotropic hormone modulating reproductive function. *Proceedings of the National Academy of Sciences, USA*, **88**, 4508–12.

Loxley, H. D., Cowell, A. M., Flower, R. J. & Buckingham, J. C. (1993). Modulation of the hypothalamo-pituitary-adrenocortical responses to cytokines in the rat by lipocortin 1 and glucocorticoids: a role for lipocortin 1 in the feedback inhibition of CRF-41 release? *Neuroendocrinology*, **57**, 801–14.

Martinez de la Escudera, G., Choi, A. L. H. & Weiner, R. I. (1994). Biphasis gabaergic regulation of GnRH secretion in GT1 cell lines. *Neuroendocrinology*, **59**, 420–5.

Mason, H. D., Margara, R., Winston, R. M., Seppala, M., Koistinen, R. & Franks, S. (1993). Insulin-like growth factor-I (IGF-1) inhibits production of IGF-binding protein-1 while stimulating oestradiol secretion in granulosa cells from normal and polycystic human ovaries. *Journal of Clinical Endocrinology and Metabolism*, **76**, 1275–9.

Mattern, L. G., Helmreich, D. L. & Cameron, J. L. (1993). Diurnal pattern of pulsatile luteinizing hormone and testosterone secretion in adult male rhesus monkeys (*Macaca mulatta*): influence of the timing of daily meal intake. *Endocrinology*, **132**, 1044–54.

Meeker, R. B., Greenwood, R. S., Hayward, J. N. (1994). Glutamate receptors in the rat hypothalamus and pituitary. *Endocrinology*, **134**, 621–9.

Mellon, P. L. (1990). Immortalization of hypothalamic gonadotropin hormone-releasing hormone (GnRH) neurons by genetically targeted tumorigenesis. *Neuron*, **5**, 1–10.

Mellon, P. L., Windle, J. J., Goldsmith, P. C., Padula, C. A., Roberts, J. L. & Weiner, R. I. (1990). Immortalization of hypothalamic gonadotropin hormone-releasing hormone (GnRH) neurons by genetically targeted tumorigenesis. *Neuron*, **5**, 1–10.

Merchenthaler, I., Lopez, F. J. & Negro-Vilar, A. (1990). Colocalization of galanin and luteinizing hormone-releasing hormone in a subset of preoptic hypothalamic neurons: anatomical and functional correlates. *Proceedings of the National Academy of Sciences, USA*, **87**, 6326–30.

Mikines, K. J., Sonne, B., Farrell, P. A., Tronier, B. & Galbo, H. (1989). Effect of training on the dose-response relationship for insulin action in men. *Journal of Applied Physiology*, **66**, 695–703.

Moenter, S. M., Caraty, A., Locatelli, A. & Karsch, F. J. (1991). Pattern of GnRH secretion leading up to ovulation in the ewe: existence of a preovulatory surge. *Endocrinology*, **129**, 1175–85.

Morley, J. P., Whitfield, J. F., Vanderhyden, B. C., Tsang, B. K. & Schwartz, J. L. (1992). A new, nongenomic estrogen action: the rapid release of intracellular calcium. *Endocrinology*, **131**, 1305–12.

Naftolin, F., Leranth, C., Perez, J. & Garcia-Segura, L. M. (1993). Oestrogen induces synpatic plasticity in adult primate neurons. *Neuroendocrinology*, **57**, 935–9.

Nyberg, C. L., Hiney, J. K., Minks, J. B. & Les Dees, W. (1993). Ethanol alters *N*-methyl-DL- aspartic acid-induced secretion of luteinizing hormone releasing hormone and the onset of puberty in the female rat. *Neuroendocrinology*, **57**, 863–8.

O'Byrne, K. T., Thalabard, J. C., Grosser, P. M., Wilson, R. C., Williams, C. L., Chen, M. D., Ladendorf, D., Hotchkiss, J. & Knobil, E. (1991). Radiotelemetric monitoring of hypothalamic gonadotropin-releasing hormone pulse generator activity throughout the menstrual cycle of the rhesus monkey. *Endocrinology*, **129**, 1207–14.

O'Conner, J. L., Wade, M. F., Brann, D. W. & Mahesh, V. B. (1993). Direct anterior pituitary modulation of gonadotropin secretion by neuropeptide Y: role of gonadal steroids. *Neuroendocrinology*, **58**, 129–35.

Okada, K., Sugihara, H., Minami, S. & Wakabayashi, I. (1993). Effect of parenteral administration of selected nutrients and central injection of g-globulin from antiserum to neuropeptide Y on growth hormone secretory pattern in food-deprived rats. *Neuroendocrinology*, **57**, 678–86.

Olster, D. H. & Ferin, M. (1987). Corticotropin-releasing hormone inhibits gonadotropin secretion in the ovariectomized monkey. *Journal of Clinical Endocrinology and Metabolism*, **65**, 262–7.

Papandreou, M. J., Asteria, C., Pettersson, K., Ronin, C. & Beck-Peccoz, P. (1993). Concanavalin A affinity chromatography of human serum gonadotropins: evidence for changes of carbohydrate structure in different clinical conditions. *Journal of Clinical Endocrinology and Metabolism*, **76**, 1008–13.

Parfitt, D. B., Church, K. R. & Cameron, J. L. (1991). Restoration of pulsatile luteinizing hormone secretion after fasting in rhesus monkey (*Macaca mulatta*): dependence on size of the refeed meal. *Endocrinology*, **129**, 749–56.

Petraglia, F., Vale, W. & Rivier, C. (1986). Opioids act centrally to modulate stress-induced decrease in luteinizing hormone in the rat. *Endocrinology*, **119**, 2445–50.

Phipps, W. R., Martini, M. C., Lampe, J. W., Slavin, J. L. & Kurzer, M. S. (1993). Effect of flax seed ingestion on the menstrual cycle. *Journal of Clinical Endocrinology and Metabolism*, **77**, 1215–19.

Ping, L., Mahesh, V. B., Wiedmier, V. T. & Brann, D. W. (1994). Release of glutamate and aspartate from the preoptic area during the progesterone-induced LH surge: *in vivo* microdialysis studies. *Neuroendocrinology*, **59**, 318–24.

Plant, T. M., Gay, V. L., Marshall, G. R. & Arslan, M. (1989). Puberty in monkeys is triggered by chemical stimulation of the hypothalamus. *Proceedings of the National Academy of Sciences, USA*, **86**, 2506–10.

Pohl, C. R., Lee, L. R. & Smith, M. S. (1989). Qualitative changes in luteinizing hormone and prolactin responses to *N*-methyl-aspartic acid during lactation in the rat. *Endocrinology*, **124**, 1905–11.

Prasad, B. M., Conover, C. D., Sarkar, D. K., Rabii, J. & Advis, J. P. (1993). Feed restriction in prepubertal lambs: effect on puberty onset and on *in vivo* release of luteinizing-hormone-releasing hormone, neuropeptide Y and beta-endorphin from the posterior–lateral median eminence. *Neuroendocrinology*, **57**, 1171–81.

Rasmussen, D. D. (1993). Episodic gonadotropin-releasing hormone release from the rat isolated median eminence *in vitro*. *Neuroendocrinology*, **58**, 511–18.

Rivest, S., Lee, S., Attardi, B. & Rivier, C. (1993*a*). The chronic intracerebroventricular infusion of interleukin-1 beta alters the activity of the hypothalamic-pituitary–gonadal axis of cycling rats. I. Effect on LHRH and gonadotropin biosynthesis and secretion. *Endocrinology*, **133**, 2424–30.

Rivest, S., Plotsky, P. M. & Rivier, C. (1993*b*). CRF alters the infundibular LHRH secretory system from the medial preoptic area of female rats: possible involvement of opoid receptors. *Neuroendocrinology*, **57**, 236–46.

Rivier, C. & Erickson, G. (1993). The chronic intracerebroventricular infusion of interleukin-1 beta alters the activity of the hypothalamic-pituitary-gonadal axis of cycling rats. II. Induction of pseudopregnant-like corpora lutea. *Endocrinology*, **133**, 2431–6.

Rossmanith, W. G., Reichelt, C. & Scherbaum, W. A. (1994). Neuroendocrinology of aging in humans: attenuated sensitivity to sex steroid feedback in elderly postmenopausal women. *Neuroendocrinology*, **59**, 355–62.

Schreihofer, D. A., Golden, G. A. & Cameron, J. L. (1993*a*). Cholecystokinin (CCK)-induced stimulation of luteinizing hormone (LH) secretion in adult male rhesus monkeys: examination of the role of CCK in nutritional regulation of LH secretion. *Endocrinology*, **132**, 1553–60.

Schreihofer, D. A., Amico, J. A. & Cameron, J. L. (1993*b*). Reversal of fasting-induced suppression of luteinizing hormone (LH) secretion in male rhesus monkeys by intragastric nutrient infusion: evidence for rapid stimulation of LH by nutritional signals. *Endocrinology*, **132**, 1879–90.

Schreihofer, D. A., Parfitt, D. B. & Cameron, J. L. (1993*c*). Suppression of luteinizing hormone secretion during short-term fasting in male rhesus monkey: the role of metabolic versus stress signals. *Endocrinology*, **132**, 1879–80.

Schweiger, U., Tuschl, R. J., Platte, P., Brooks, A., Laessle, R. G. & Pirke, K. M. (1992). Everyday eating behavior and menstrual function in young women. *Fertility and Sterility*, **57**, 771–5.

Scott, C. J. & Clarke, I. J. (1993). Evidence that changes in the function of the subtypes of the receptors for gamma-amino butyric acid may be involved in the seasonal changes in the negative-feedback effects of oestrogen on gonadotropin-releasing hormone secretion and plasma leutinizing hormone levels in the ewe. *Endocrinology*, **133**, 2904–12.

Segaloff, D. L. & Ascoli, M. (1993). The lutotropin/choriogonadotropin receptor . . . 4 years later. *Endocrine Review*, **14**, 324–47.

Selvais, P. L., Colin, I. M., Adam, E., Kasa-Vubu, J. Z., Denef, J. F. & Maiter, J. M. (1993). Effects of hypophysectomy on galaninergic neurons in the rat hypothalamus. *Neuroendocrinology*, **58**, 539–47.

Shalts, E., Xia, L., Xiao, E. & Ferin, M. (1994). Inhibitory effect of arginine-vasopressin on LH secretion in the ovariectomized rhesus monkey. *Neuroendocrinology*, **59**, 336–42.

Sherwood, N. M., Lovejoy, D. A. & Coe, I. R. (1993). Origin of mammalian gonadotropin-releasing hormone. *Endocrine Reviews*, **14**, 241–54.

Shughue, P. J. & Dorsa, D. M. (1993). Oestrogen modulates the growth-associated protein GAP-43 (neuromodulin) mRAN in the rat preoptic area and basal hypothalamus. *Neuroendocrinology*, **57**, 439–47.

Spergel, D. J., Krsmanovic, L. Z., Stojilkovic & Catt, K. J. (1994). Glutamate modulates $[Ca^{2+}-]i$ and gonadotropin-releasing hormone secretion in immortalized hypothalamic GT1-7 neurons. *Neuroendocrinology*, **59**, 309–17.

Stanton, P. G., Pozvek, G., Burgon, P. G., Robertson, D. M. & Hearn, M. T. W. (1993). Isolation and characterization of human LH isoforms. *Journal of Endocrinology*, **138**, 529–43.

Thind, K. K., Boggan, J. E. & Goldsmith, P. C. (1993). Neuropetide Y system of the female monkey hypothalamus: retrograde tracing and immunostaining. *Neuroendocrinology*, **57**, 289–98.

Walberg, R. J., Franke, W. D. & Gwazdauskas, F. C. (1992). Response of beta-endorphin and oestradiol to resistance exercise in females during energy balance and energy restriction. *International Journal of Sports*, **13**, 542–7.

Weiss, J., Axelrod, L., Whitcomb, R. W., Harris, P. E., Crowley, W. F. & Jameson, J. L. (1992). Hypogonadism caused by a single amino acid substitution in the beat subunit of luteinizing hormone. *New England Journal of Medicine*, **326**, 179–83.

Wide, L. & Bakkos, O. (1993). More basic forms of both human follicle-stimulating hormone and luteinizing hormone in serum at midcycle compared with the follicular or luteal phase. *Journal of Clinical Endocrinology and Metabolism*, **76**, 885–9.

Wilding, J. P. H., Lambert, P. D., Al-Dokhayel, A. A. M., Gilbey, S. C., Ghatei, M. A. & Bloom, S. R. (1993). Evidence for NPY-opioid interactions in the control of food intake. Abstract #0C29.

Williams, C. L., Nishihara, M., Thalabard, J. C., O' Byrne, K. T., Grosser, P. M., Hotchkiss, J., Knobil, E. (1990a). Duration and frequency of multiunit electrical activity associated with the hypothalamic gonadotropin releasing hormone (GnRH) pulse generator in the Rhesus monkey: differential effects of morphine. *Neuroendocrinology*, **52**, 225.

Williams, C. L., Nishihara, M., Thalabard, J. C., Grosser, P. M., Hotchkiss, J., Knobil, E. (1990b). Corticotropin releasing factor and gonadotropin-releasing hormone pulse generator activity in the Rhesus monkey. *Neuroendocrinology*, **52**, 133.

Wilson, R. C., Kesner, J. S., Kaufman, J. M., Uemura, T., Akema, T., Knobil, E. (1984). Central electrophysiologic correlates of pulsatile luteinizing hormone secretion in the Rhesus monkey. *Neuroendocrinology*, **39**, 256.

Wynick, D., Hammond, P. J., Akinsanya, K. O. & Bloom, S. R. (1993). Galanin regulates basal and oestrogen-stimulated lactotroph function. *Nature*, **364**, 529.

3 Lactation, condition and sociality: constraints on fertility of non-human mammals

P. C. LEE

Introduction

A major focus of research in mammalian behavioural ecology has been on the factors controlling reproductive success among different individuals (e.g. Clutton-Brock, 1988). This concept of 'reproductive success', defined as the number of offspring who survive to breeding age, is central, in that it lies at the heart of explanations about why individuals vary in their behaviour associated with courtship, access to mates and parenting, and ultimately influences patterns of mating systems and sociality. Reproductive success is thus a measure of individual fitness, containing a number of components and allowing for the assessment of sources of variance. The main elements contributing to reproductive success are age-specific fertility and fecundity, offspring survival and breeding lifespan (Table 3.1). As such, an understanding of the constraints on fertility is crucial to building a complete picture of behaviour in relation to reproduction, and in descriptions of individual variation and its sources.

However, mammalian field studies tend to be observational rather than invasive, and only a limited understanding of fertility mechanisms exists. The majority of studies produce correlational findings and causal links remain speculative. Behavioural physiology on wild species is in its infancy, and of necessity, we must draw on captive or domestic species in order to gain a broader perspective on fertility mechanisms. This aside, a sufficient number of studies on reproductive variation exist for some general observations to be made about the links between individual characteristics, behaviour and fertility. The major factors highlighted as influencing fertility among non-human mammals are, unsurprisingly, similar to those seen in humans. Lactation, its energy costs, the energy balance of the female, her overall body condition and her social context are all proposed as major influences on fertility in mammals. Among domestic species,

25

Table 3.1. *Components of reproductive success in mammalian species (after Clutton-Brock, 1988), indicating the potential ecological and social sources of variance*

Component 1: *Survival to breeding age*

Sources of variance: Maternal condition
Infant growth rates
Extrinsic sources of mortality (predation, disease)

Component 2: *Reproductive lifespan (L)*

Sources of variance: Age at first reproduction
Age-specific mortality
Maximum lifespan

Component 3: *Fecundity per year over breeding lifespan (F)*

Sources of variance: Environmental constraints on fertility
Social constraints on fertility and fecundity
Lactational constraints on fertility
Genetic variation in fertility

Component 4: *Offspring survival between birth and reproduction (S)*

Sources of variance: Maternal condition and maintenance of growth
Extrinsic sources of mortality

---

genetic variation is also known to affect fertility. And finally, there is considerable variance in breeding success between individual males within a species, which may play a further role in the successful breeding of females. The evidence for the effects of these factors on reproduction will be reviewed briefly below.

Lactational constraints

The role of lactation in fertility among non-human mammals has long been noted for domestic species; however, cattle can be mated and reconceive during lactation, while pigs remain anoestrus throughout (Cowie & Buttle, 1980). The underlying hormonal mechanisms for these species differences are still under investigation. There is some evidence from sheep (Arnold, Wallace & Maller, 1979) that, once a critical threshold is reached for the mother in her relations between energy intake and milk output (see also Lee, Majluf & Gordon, 1991), an abrupt weaning occurs, and oestrus returns.

The first evidence that lactation played a major role in the fertility of

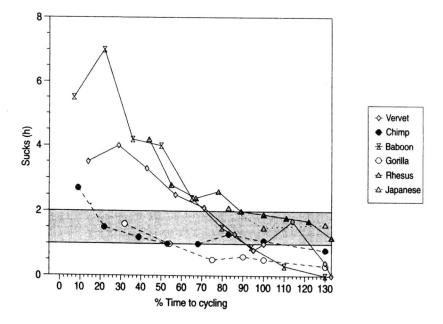

Fig. 3.1. Suckling bout frequencies in primate species plotted as a function of the relative time to the resumption of cycling, in order to facilitate comparisons between species where interbirth intervals range from 1–5 years. The threshold area of 1–2 bouts/h is indicated, showing the general pattern of a drop in frequencies either prior or at the average onset of cycling (100%). Data from: Vervets (Lee, 1987), rhesus (Gomendio, 1989a), Japanese macaques (Tanaka *et al.*, 1993), chimpanzees (Clark, 1977), gorillas (Stewart, 1988) and baboons (Nicolson, 1982). In the case of the baboons, suckling frequencies were halved for the first 3 intervals to ensure comparability with the other studies, where a 60 s bout criterion was used rather than 30 s in the case of the baboons.

non-human primates came from the observations that the death of the infant shortened interbirth intervals in langurs (Hrdy & Hrdy, 1976; Winkler, Loch & Vogel, 1984), macaques (Hadidian & Bernstein, 1979), baboons (Altmann, Altmann & Hausfater, 1978; Dunbar, 1980a) and apes (Goodall, 1983). In these species, the interbirth intervals were typically two years or longer, and an infant's death in its first year of life led to a rapid subsequent reconception.

These observations were subsequently linked to the mechanism of suckling frequencies for a number of species (baboons, Nicolson, 1982; gorillas, Stewart, 1988; vervets, Lee, 1987; rhesus, Gomendio, 1989a). High suckling frequencies were associated with a lack of menstrual cycling, and only as the suckling frequencies declined did cycling resume and the probability of conception increase (Fig. 3.1). Even in seasonally breeding species, high suckling frequencies during the breeding season resulted either in the complete absence of conception or in a substantial delay to conception within the season (Gomendio, 1989b; Johnson, Berman & Malik, 1993).

The hormonal profiles of female primates showed a similar trend to that observed in lactating humans. Frequent suckling has been related to a suppression of GnRH, at least for macaques (Gordon, 1981; Pope, Gordon & Wilson, 1986; Wilson et al., 1988). It thus appears that the underlying mechanisms relating lactation to fertility among primates are similar to those in humans (see Gomendio, 1989b). When ovulation returns, and reconception results, lactation does not terminate and pregnancy is compatible with extended lactation in non-human primates, just as it is in humans. Milk production can continue for up to 100 out of the 173 days of gestation in Japanese macaques. This was found to be especially marked among females with shorter interbirth intervals, suggesting that milk remains a critical nutritional element of the diet for at least the first 18 months in this species (Tanaka, Hayama & Nigi, 1993).

Among a number of other wild mammals, lactation can again be linked to a delay in reconception. For example, in elephants, the resumption of oestrus was shown to be related to a drop in suckling frequencies (Lee, 1986; Lee & Moss, 1986), even when the total lactation period was of the order of 5 years. For lions, the killing of the young cubs by rival invading males produces an immediate oestrus in the females (Packer et al., 1988). Termination of lactation through infanticide is a competitive reproductive strategy seen in a number of primate (Struhsaker & Leyland, 1987) and other species, from rodents to red deer (Bartos & Madlafousek, 1994). At this stage, we can only speculate that the hormonal controls of the lactational anoestrus will be similar to those observed in humans and primates, as hormonal profiles for elephants, lions or other wild mammals are poorly understood and difficult to assess.

Social constraints

The small New World species of primates, the marmosets and tamarin monkeys, are unusual in that they tend to have a post-partum oestrus and reconceive while lactating at high frequencies. They are capable of producing twins at 5–6 m intervals in captivity (Rothe, 1975). Recent detailed studies of their reproductive hormones has highlighted the importance of the social, as opposed to the lactational, context for these animals (for review see Abbott, 1992). When two or more adult females are housed together, typically only a single dominant female will show oestrous cycling. The other females are hormonally suppressed with a complete absence of LH surge leading to ovulation. Similar results have been found for some of the larger social species of primates, such as talapoin monkeys (Bowman, Diller & Keverne, 1978), leading to the formulation of theories as to the social suppression of reproduction.

The phenomenon of social suppression of oestrus cycling in the primates has received considerable attention, and a number of other observations of reduced fertility in the context of social stress have been made (see Dunbar, 1990). Aggression towards subordinate females on the part of reproductively active dominant individuals has been postulated as the main mechanism interfering with cycling through the adrenal – hypothalamus link (primates, Abbott, 1992; pigs, Mendl, Zanella & Broom, 1992). Subordinate female baboons, who experienced high rates of aggression, had longer follicular phases to their cycles (Rowell, 1970), which could reduce the likelihood of a conception. Social suppression of reproduction has long been noted in small carnivores such as mongooses (Rood, 1980), large species such as wolves, jackals (MacDonald & Moehlman, 1982) and African wild dogs (Frame *et al.*, 1979), as well as primates such as gelada baboons (Dunbar, 1980*b*). Among dwarf mongoose (Creel *et al.*, 1992), ovulatory suppression is incomplete, but low baseline oestrogen levels in combination with low mating frequencies tend to result in lower conception success.

Direct links between social stress and fecundity have also been postulated. Among baboons and rhesus monkeys, attacks on cycling and gestating subordinate females are proposed to result in early abortions (Sackett *et al.*, 1974; Wasser & Starling, 1986). Reduced fecundity through induced abortion appears to be especially marked for wild horses, where spontaneous abortion rate is about 10%, and higher rates can be socially induced through aggression directed to subordinate mares (Berger, 1986). The importance of individual status within the social group in determining variation in reproduction is demonstrated further in suggestions that the fertility of the subordinates can be limited when access to fertile males is constrained through inter-female aggression (Niemeyer & Anderson, 1983; Bruce & Estep, 1992).

Some other interesting potential effects of the social context on fertility have been postulated. Colmenares and Gomendio (1988) found that the introduction of a novel adult male initiated cycling among baboons females, even for those in early lactation. Furthermore, the presence of the novel male effectively advanced the age of first cycling in adolescent females. High levels of arousal in relation to social interactions have been

Table 3.2. *Social influences on fertility in mammals*

Social suppression of fertility
Causal factors
High levels of aggression directed towards subordinate individuals
Stress resulting from sustained low levels of inter-individual conflict
Lower condition resulting from competition over foods

Effects on subordinate females

Effects	Species
Loss or insufficiencies of ovulatory cycles	Marmosets, tamarins, talapoins
	Wolves, jackal, mongoose
	Naked mole rat
Delayed first conception	Macaques
Reduced fertility	Baboons, macaques, vervets
Increased infant mortality	Baboons, macaques
	Wild horses

Effects on subordinate males

Effects	Species
Reduced sexual activity	Talapoin, squirrel monkey,
	Hamadryas baboons
Reduced testosterone levels	Elephants
	dwarf lemurs
Increased cortisol levels	Baboons
Reduced access to mates	Most polygynous species

Social enhancement of fertility
Causal variables
Novelty in sexual partners

Effects	Species
Earlier age of first cycling	Baboons
Resumption of cycles when lactating	
Early weaning of infants	

Male presence/advertisement/mating activity

Effects	Species
Advance oestrus	Red deer
Synchronize oestrus	Lions
	Elephants

suggested to be influential in the fertility of other wild mammals. The roaring of a rutting stag has been shown to induce oestrus among female red deer (McComb, 1987), while the presence of one oestrous female and her consequent mating activities are thought to initiate and synchronize oestrus among lion females (e.g. Packer *et al.*, 1988) and elephants (Moss & Dobson, in prep) (Table 3.2).

Condition effects

For many non-primate species, a post-partum oestrus is observed, with a high probability of conception. However, the probability of ovulation, even during the post-partum oestrus is still related to the intensity and costs of lactation, at least in some domestic species (Jainudeen & Hafez, 1980*a*). Observations on many wild species have also suggested that the energy budget of the female, especially during lactation, determines her body condition and subsequent probability of conception (e.g. red deer, Clutton-Brock *et al.*, 1982; Clutton-Brock, Albon & Guinness, 1989; wild horses, Rubenstein, 1986; zebras, Ginsberg, 1989; Becker & Ginsberg, 1990; cheetahs, Laurenson, 1992) (see Table 3.3 for examples of ecological effects on female reproductive potential). Depletion of body fat or muscle mass during lactation, producing a reduced weight for height/body length, can indicate poor condition and reduced fertility. It has been speculated that poor condition leads to higher stress levels, producing the reduction in the frequency or indeed the existence of oestrus during poor seasons or hard years (Follett, 1984). However, condition is notoriously difficult to assess in living wild mammals. Recent work has used indices such as limb circumference and skinfold fat (Bowman, 1991) and blood parameters (haemoglobin concentrations, cell volume, white cell parameters; Laurenson, 1992) to assess condition depletion during lactation.

Among the primates, a substantial body of indirect evidence suggests that general female condition affects fertility. In poor habitats, age at first reproduction and interbirth intervals are extended among vervet monkeys (Lee, 1984, 1987; Cheney *et al.*, 1988), while for baboons, inter-annual variation in food availability also affects growth rates and age at first reproduction (Strum, 1991). In baboons and Barbary macaques, resource depletion due to competition with other grazing species prolongs interbirth intervals (Strum & Western, 1982; Drucker, 1984). Reductions of food availability in the form of the withdrawal of supplemental foods obtained through provisioning also markedly decreases the reproductive rate of female macaques (Mori, 1979; Sugiyama & Ohsawa, 1982; Malik & Southwick, 1988; Loy, 1988). A general index of habitat quality, such as rainfall, accounts for a substantial proportion of the variation in population fecundity among baboons and macaques (see Dunbar, 1992). Bercovitch and Strum (1993) have suggested that GnRH insufficiencies among cycling females result from the habitat in two ways; firstly through food intake, in the form of nutrients and quantity, and secondly through the metabolic costs associated with longer foraging distances in poor seasons or years.

Table 3.3. *Examples of ecological sources of demographic and reproductive variation for different mammals*

Primates	
Cause: Lower food availability	
Variable	Relative change
Age at first reproduction	20% decrease in breeding lifespan
Birth rate (per Female)	67% decrease in fecundity
Survivorship of infants	60% reduction
Population growth rate	15%/yr to −5%/yr (extinction!)
Red Deer	
Cause: Population density increase	
Variable	Relative change
Calf birth weight	20% decrease
Calf winter survival	80% reduction (low birth weight)
Lions	
Cause: Dry season food limitation	
Variable	Relative change
Cub survival	45% decrease over wet season
Pride size	66% decrease in RS in very small/very large prides
Elephants	
Source: Drought years	
Variable	Relative change
Conceptions	100% cessation of oestrus
Interbirth intervals	47% increase
Calf mortality	50% increase
Zebras	
Source: Water shortage during lactation	
Variable	Relative change
Weaning age	50% decrease in duration of lactation
Foraging time costs	45% increase in maternal time budget

Some evidence for a direct link between condition and fertility exists for domestic species. Cattle, sheep and horses all show a reduction in fertility as a result of undernutrition or vitamin deficiencies (Jainudeen & Hafez, 1980a). The physiological mechanism is thought to be that of gonadotropin insufficiencies, resulting in failure of the follicular phase. The effect of poor condition is greatest for younger, growing females.

Ovulation rate correlates with weight at breeding for pigs (Anderson, 1980). However, as is true for humans, reductions in fecundity in relation to energy deficit are most pronounced when the energy reduction is severe (at

50% of normal), and sustained, again probably mediated through the overall condition of the female during successive pregnancies. (cats, rats, pigs, Anderson, 1980). Captive macaques fed on low energy and protein diets also have reduced fecundity (Riopelle & Favret, 1977), while well-fed counterparts have higher reproductive rates than are observed in the wild (Loy, 1988; Lyles & Dobson, 1988; Dunbar, 1990). Among sheep and goats, environmental stress, such as temperature or low nutrition, reduces fertility, while the rapid onset of oestrus is associated with seasonal grass productivity (Hulet & Shelton, 1980). In species such as wildebeest and buffalo, a similar response to seasonal increases in primary productivity has been observed. Among wildebeest, the rut occurs after the rains (which determine grass production), when females are in peak condition (Estes & Estes, 1979). For African buffalo, the timing of the rains again determines when both females and males attain peak condition, which then determines mating (Sinclair, 1977; Prins, 1987). Elephants also show an extremely reduced rate of oestrus during droughts, and females are more likely to come into oestrus one to two months after the rains in normal seasonal cycles (Poole, 1987; Moss & Poole, 1983). Despite their large size, body reserves may not be sufficient to sustain prolonged lactation and reconception during seasonal or other periods of food shortage (Lee & Moss, 1986).

Many northern temperate ungulates reproduce in the fall after the relatively long period of summer grass growth, which allows for weight gain. Females thus enter pregnancy at peak condition and weight, producing infants in the following spring. Among mountain sheep (Fiesta-Bianchet, 1988a; Berger, 1991) pregnant females preferentially seek out resources of higher quality. Over-winter weight loss during gestation is widespread among wild mammals (deer, Torbit *et al.*, 1985; Clutton-Brock *et al.*, 1989; polar bears, Ramsay & Stirling, 1988). The timing of seasonal breeding in northern species is probably influenced less by the current condition of the females than by the costs of gestation during a period of low food, and the necessity to calve when food is again abundant. Energy costs of gestation and lactation are balanced against the female's body condition at the time of conception, with consequences for seasonal breeding and reproductive performance.

In some domestic species, age is also a factor, in that older females in species such as pigs have a higher ovulation rate than do younger females, but as is typical in most mammals, ovulation rate declines at old ages. The confounding factor is that weight of females also increase with subsequent pregnancies, so that older females are heavier than are primiparous females (pigs, Anderson, 1980). The trade-off between growth and reproduction is an important consideration, which may affect fertility in the younger, still growing, age classes most strongly (e.g. Bowman, 1991). Among a number of primate species, slow-growing females reach adolescence later, and have reduced success in their first breeding attempts (e.g. baboons, Bercovitch & Strum, 1993; macaques, Bercovitch & Berard, 1993).

A second trade-off, that between lactation and condition, has been demonstrated for red deer, where mothers who sustain high lactational costs are less likely to reconceive (Clutton-Brock, Guinness & Albon, 1983; Loudon & Kay, 1983). Among mice (Konig & Markl, 1987) and rats (Smith, 1991), mothers who reconceive at the post-partum oestrus had longer subsequent gestations, smaller pups in the second litter, and low growth rates among pups of the first litter. Small species with short life spans and large litters, and which typically experience environmental unpredictability, minimise both the duration of care and weight gain of offspring when confronted by energetic constraints (Konig & Markl, 1987). Among rhesus macaques, mothers who sustained high infant growth rates during lactation also were less likely to reconceive (Bowman & Lee, 1995). Bonnet macaque mothers who are made to work for food during lactation have less secure relationships with their infants (Andrews & Rosenblum, 1988), which could potentially affect fertility if suckling frequencies are altered. For wild females, the costs of energy loading on mothers during foraging can be substantial (Dunbar & Dunbar, 1988; Altmann & Samuels, 1992) and females are only likely to reconceive after regaining a threshold weight (Berkovitch, 1987). Among marmosets and tamarins, reproductive rates and infant survival are higher when females have access to caretakers (Goldizen, 1987; Baker, Dietz & Kleiman, 1993), suggesting that social factors can influence the allocation of lactational energy, with concomitant consequences for subsequent reproductive events.

A further source of fertility variation due to offspring sex has been noted. Sons have higher growth rates, suckle at higher frequencies and are more demanding than are daughters (Clutton-Brock, 1991). Thus, suckling a son among sexually dimorphic species appears to be more costly physiologically; sons suckle more frequently (and thus there is the potential for lactational inhibition of oestrus), and for longer, leading to a greater condition depletion. Red deer yeld hinds, who delay reproduction for a year and are thus likely to be in good condition, are more frequently the mothers of sons than are hinds who reproduce in consecutive years (Clutton-Brock et al., 1983, 1989), suggesting that maternal condition may be both a function of the sex of her previous offspring and bias the sex of her subsequent offspring. In general, rearing a son produces longer interbirth intervals (red deer, Clutton-Brock et al., 1983; elephants, Lee & Moss, 1986; Indian cattle, Dhillon et al., 1970; bison, Wolf, 1988; mountain sheep,

Fiesta-Bianchet, 1988*b*), greater maternal weight loss during lactation (reindeer, Kojola & Eloranta, 1989), and potentially, reduced maternal survival.

Condition, diet and growth in early life have also been related to reproductive performance among female baboons (Altmann, 1991). Poor early diets, leading to slower growth and reduced general levels of fatness (Altmann *et al.*, 1993) result in poor reproductive performance; such females have longer interbirth intervals, and greater numbers of cycles prior to conception, as well as a late age of reproductive maturity. Well-fed baboons have short interbirth intervals, and levels of body fat which approach those seen in captive species. The role of body fat in reproduction among wild primates is difficult to assess directly, as such measures are generally not available. Where body composition is known, free-ranging, wild baboons had less than 2% body fat (Altmann *et al.*, 1993). However, among captive species the proportion of mass which is fat is generally great (over 15%, see Altmann *et al.*, 1993), and thus conclusions about diet, energy and body composition affecting fertility are probably biased in this respect. In some ways, captive fat primate females are probably more similar to western well-nourished women than they are to their wild counterparts.

There is an interaction between the social and condition variables, such that in species where dominance relations are observed, individuals of lower status tend to be those who have reduced access to resources, at least among primates (Dittus, 1979; Whitten, 1983). As a consequence, their fertility may be impaired (Harcourt, 1987). In this case, the mechanism by which a reduction in fertility occurs is that of social prevention of the intake of adequate resources, but the underlying relationship is still probably condition related. This example demonstrates the difficulty of separating interacting variables, even in relatively simple, non-cultural, systems seen in non-humans (Table 3.4).

Male effects

Do males play a role in the fertility of females? Data on sperm competition from a number of mammals suggests that multiple copulations may *reduce* fertility. Among rodents, interactions between sperm in multiply-mated females produce an effect of first mating having priority of fertilization, unless the delay between copulations is substantial. In such cases, the last male to mate may be the one fertilizing the egg (Dewsbury, 1984). Observations on chimpanzees suggest that unique consortships persisting

Table 3.4. *A summary of the relations between condition and fertility as well as some of the intervening factors*

I. Seasonal food cycles
Source: Environmental variation due to rainfall, temperature on daylength

---

a) Seasonal attainment of minimum weight, fat or weight for size
b) Onset of hormonal function (condition, photoperiod)
c) Selection for birth periods to maximize energy intake during lactation

Effect: Seasonal cycling and breeding

II. Food restriction during lactation
Source: Energy requirements increase; Energy and time costs of foraging increase;
Changes in food selection;
Socially restricted intake and high rates of received aggression

---

a) Reduced immunological and ovulatory function
b) Loss of body weight/fat
c) Reduced infant birth weight and growth
d) High suckling frequencies, especially among male offspring in sexually dimorphic species

Effects: Delay in subsequent reproduction
Increase in infant mortality
High variance in weaning age

III. Food restriction during early growth
Source: Cohort or year of birth; Episodes of extreme environmental variation;
Social limits to nutrition; Poor maternal condition or inability to invest

---

a) Slow growth throughout immaturity
b) Low body size attained at reproductive age

Effects: Later age at first reproduction
Reduced reproductive performance over the lifespan
Increased mortality during reproduction

---

over a number of days produce a much higher rate of conception than do multiple sequential, promiscuous matings, despite the social risks of aggression directed at the consorting pair (Hasegawa & Hiraiwa-Hasegawa, 1983). For chimpanzees, either immunological interactions between the sperm, or social stresses associated with promiscuous mating, can reduce the potential for fertilization. Among lions, sequential copulations at frequent intervals over several days appear to be necessary either to induce ovulation in the oestrus female or to provide sufficient sperm numbers, or both. Interruption of the copulatory relationship decreases fertility.

Malformations of the sperm, lack of sperm viability, and reduced sperm numbers all contribute to low male and female fertility in domestic mammals (Jainudeen & Hafez, 1980b), and these are thought to play a role in some wild species (e.g. cheetah, Caro *et al.*, 1987). Chronic energy shortage also affects males, through a later age at first reproduction, where undernutrition suppresses testicular endocrine activity. However, under or over feeding may have a greater effect on the desire to mate among adult males. Among elephants, where reproduction among bulls is associated with a physiological state called musth, bulls fail to enter musth or to exhibit normal musth cycles during droughts (Poole, 1987). As musth period entails a considerable condition loss (Poole, 1987), it appears that males are unable to initiate or sustain this energetically demanding reproductive phase when they are in poor condition. This was also demonstrated when bulls who were injured in the previous year's musth period failed to enter musth in the following year (Poole, 1989). Similar costs of reproductive activity on male survival and lifespan have been observed in wild ungulates and species as diverse as fruit flies (Partridge, 1988).

Other important constraints on male fertility appear to be a consequence of their early environment. Birth outside the optimal seasonal period, mothers unable to invest sufficiently either due to their own physical condition or to prevailing environmental conditions, physical or social stress during infancy and adolescence are all important in the future survival and reproduction of males (red deer, Clutton-Brock *et al.*, 1988), just as among females. In the red deer, cohorts of poor reproducers have been observed, suggesting that the effects of early environmental stresses persist throughout the males' lives.

Genetic effects

There are few data on genetic variability in wild mammals, and none on genetic factors in fertility. However, fertility variation, at least in well-studied domestic species, can be accounted for by genetic variables; 40% of the variance in fertility among pigs was due to breeding (Anderson, 1980). Among wild species, even data on familial histories are difficult to obtain, since studies of reproduction over the lifespan of several generations of females are rare. Daughters of dominant Japanese macaque females, with high reproductive rates, have been shown to reproduce at a younger age and more frequently than their subordinate peers (Mori, 1979), but whether this is due to any genetic difference or simply to the effect of early nutrition is, as yet, unknown.

Genetic effects through the paternal line have been noted for some primates, where certain males tended to sire offspring with consistently poor survivorship (Sackett, 1990). The genetic effects expressed through the father may be less related to fertility among females than to the survival of the foetus or neonate. In part, these may be the consequence of some genetic abnormality, which simply causes death. Alternatively, the male genetic effects may act through a low birth weight and poor growth of the infant (e.g. Bowman, 1991), with consequences for its subsequent reproductive performance, as well as survival potential.

Conclusions

Mammalian fertility is affected by a wide range of intrinsic and external factors, of which lactation and body condition in relation to energy available from the environment appear to be the most influential. While the suckling/fertility relations are relatively well understood in non-human primates, the precise ways in which body composition, weight and stature interact with the physical and social environment, and the genetic make-up of the individual, have yet to be defined. Most of the observations of low fertility among poorly nourished animals are correlational, although work on domestic species has suggested that body weight and nutrient intake are especially important in fertility rates. Among mammals, the importance of early nutrition and growth on subsequent reproductive potential has highlighted areas for further exploration; indeed animals models may be relevant for understanding some of these effects in humans.

Social factors are also important in mammalian reproduction. For the most part, social variables, such as relative dominance rank, differentially reduce the fertility of individuals of low dominance status, possibly acting through the effects of stress hormones on the production on patterning of reproductive hormones. Furthermore, social factors may also interact with condition in that subordinate individuals also tend to have reduced priority of access to resources and generally low competitive abilities, leading to the potential for socially mediated energy limitations and reduced condition. These social limits to nutrient and energy intake may be far more pronounced during times of environmental stress or indeed during lactation, further confounding any simple relations between habitats, condition and fertility. The effect of competitive variation in reproductive potential is especially marked for males in most species of polygynously breeding mammals. Finally, genetic make-up, studied among livestock, also remains a considerable potential source of variance which has yet to be determined for most mammalian species.

Despite abundant research on captive, domestic and wild species, a great deal of work remains to be done before the constraints and the mechanisms affecting mammalian reproduction are understood. At this stage, we benefit from the work on humans, as well as that on domestic species, to establish the mechanisms. However, the important questions of function, of the consequences of natural variation in fertility on lifetime reproductive success, can best be examined in wild mammalian species.

REFERENCES

Abbott, D. H. (1992). Social conflict and reproductive suppression in marmoset and tamarin monkeys. In *Primate Social Conflict*, ed. W. A. Mason and S. P. Mendoza, pp. 331–73. Albany, New York: SUNY Press.

Altmann, J. & Samuels, A. (1992). Costs of maternal care: infant carrying in baboons. *Behavioural Ecology and Sociobiology*, **29**, 391–8.

Altmann, J., Altmann, S. A. & Hausfater, G. (1978). Primate infant's effects on mother's future reproduction. *Science*, **201**, 1028–30.

Altmann, J., Schoeller, D., Altmann, S. A., Murithi, P. & Sapolsky, R. M. (1993). Body size and fatness of free-living baboons reflect food availability and activity levels. *American Journal of Primatology*, **30**, 149–61.

Altmann, S. A. (1991). Diets of yearling female primates (*Papio cynocephalus*) predict lifetime fitness. *Proceedings of the National Academy of Sciences*, **88**, 420–3.

Anderson, L. L. (1980). Pigs. In *Reproduction in Farm Animals*, ed. E. S. E. Hafez, pp. 358–86. Philadelphia: Lea & Febiger.

Andrews, M. W. & Rosenblum, L. A. (1988). Relationship between foraging and affiliative social referencing in primates. In *Ecology and Behaviour of Food-Enhanced Primate Groups*, ed. J. E. Fa and C. H. Southwick, pp. 247–68. New York: Alan R. Liss.

Arnold, G. W., Wallace, S. R. & Maller, R. A. (1979). Some factors involved in the natural weaning processes in sheep. *Applied Animal Ethology*, **5**, 43–50.

Baker, A. J., Dietz, J. M. & Kleiman, D. G. (1993). Behavioural evidence for monopolization of paternity in multi-male groups of golden lion tamarins. *Animal Behaviour*, **46**, 1091–103.

Bartos, L. & Madlafousek, J. (1994). Infanticide in a seasonal breeder: the case of red deer. *Animal Behaviour*, **47**, 217–19.

Becker, C. D. & Ginsberg, J. R. (1990). Mother-infant behaviour of wild Grevy's zebra: adaptations for survival in semi-desert East Africa. *Animal Behaviour*, **40**, 1111–18.

Bercovitch, F. B. (1987). Female weight and reproductive condition in a population of olive baboons (*Papio anubis*). *International Journal of Primatology*, **12**, 189–95.

Bercovitch, F. B. & Berard, J. D. (1993). Life history costs and consequences of rapid reproductive maturation in female rhesus macaques. *Behavioural Ecology and Sociobiology*, **32**, 103–9.

Bercovitch. F. B. & Strum, S. C. (1993). Dominance rank, resource availability and reproductive maturation in female savanna baboons. *Behavioural Ecology and Sociobiology*, **33**, 313–18.

Berger, J. (1986). *Wild Horses of the Great Basin*. Chicago: University of Chicago Press.

Berger, J. (1991). Pregnancy incentives, predation constraints and habitat shifts: experimental and field evidence for wild bighorn sheep. *Animal Behaviour*, **41**, 61–78.

Bowman, J. E. (1991). *Life History, Growth and Dental Development in Young Primates: A Study Using Captive Rhesus Macaques*. University of Cambridge: PhD Thesis.

Bowman, J. E. & Lee, P. C. (1995). Growth and threshold weaning weights among captive rhesus macaques. *American Journal of Physical Anthropology*, **96**, 159–75.

Bowman, L. A., Diller, S. R. & Keverne, E. B. (1978). Suppression of oestrogen-induced LH surge by social subordination in talapoin monkeys. *Nature*, **275**, 56–8.

Bruce, K. E. & Estep, D. Q. (1992). Interruption of and harassment during copulation by stumptail macaques, *Macaca arctoides*. *Animal Behaviour*, **44**, 1029–44.

Caro, T. M., Holt, M. E., FitzGibbon, C. D., Bush, M., Hawkwy, C. M. & Kock, R. A. (1987). Health of free-living cheetahs. *Journal of Zoology (London)*, **212**, 573–84.

Cheney, D. L., Seyfarth, R. M., Andelman, S. & Lee, P. C. (1988). Reproductive success in vervet monkeys. In *Reproductive Success*, ed. T. H. Clutton-Brock, pp. 384–402. Chicago: University of Chicago Press.

Clark, C. B. (1977). A preliminary report on weaning among chimpanzees of the Gombe National Park, Tanzania. In *Primate Bio-Social Development*, eds. S. Chevalier-Skolnikoff and F. F. Poirer, pp. 235–60. New York: Garland STMP Press.

Clutton-Brock, T. H. (Ed) (1988). *Reproductive Success*. Chicago: University of Chicago Press.

Clutton-Brock, T. H. (1991). *The Evolution of Parental Care*. Princeton: Princeton University Press.

Clutton-Brock, T. H., Albon, S. D. & Guinness, F. E. (1988). Reproductive success in male and female red deer. In *Reproductive Success*, ed. T. H. Clutton-Brock, pp. 325–43. Chicago: University of Chicago Press.

Clutton-Brock, T. H., Albon, S. D. & Guinness, F. E. (1989). Fitness costs of gestation and lactation in wild mammals. *Nature, London*, **337**, 260–2.

Clutton-Brock, T. H., Guinness, F. E. & Albon, S. D. (1983). The costs of reproduction to red deer hinds. *Journal of Animal Ecology*, **52**, 367–84.

Clutton-Brock, T. H., Iason, G. R., Albon, S. D. & Guinness, F. E. (1982). Effects of lactation on feeding behaviour and habitat use in wild red deer hinds. *Journal of Zoology (London)*, **198**, 227–36.

Colmenares, F. & Gomendio, M. (1988). Changes in female reproductive condition following the introduction of new males to a colony of hamadryas and hybrid baboons. *Folia Primatologia*, **50**, 157–74.

Cowie, A. T. & Buttle, H. L. (1980). Lactation. In *Reproduction in Farm Animals*, ed. E. S. E. Hafez, pp. 284–303. Philadelphia: Lea & Febiger.

Creel, S., Creel, N., Wildt, D. E. & Monfort, S. L. (1992). Behavioural and endocrine mechanisms of reproductive suppression in Serengeti dwarf mongooses. *Animal Behaviour*, **43**, 231–45.

Dewsbury, D. A. (1984). Sperm competition in muriod rodents. In *Sperm Competition and the Evolution of Animal Mating Systems*, ed. R. L. Smith, pp. 547–71. London: Academic Press.

Dhillon, J. S., Acharya, R. M., Tiwana, M. S. & Aggarwal, S. C. (1970). Factors affecting the interval between calving and conception in Hariana cattle. *Animal Production*, **12**, 81–7.

Dittus, W. J. P. (1979). The evolution of behaviours regulating density and age-specific sex ratios in a primate population. *Behaviour*, **69**, 265–302.

Drucker, G. R. (1984). The feeding ecology of the Barbary macaque and cedar forest conservation in the Moroccan Moyen Atlas. In *The Barbary Macaque: A Case Study in Conservation*, ed. J. E. Fa, pp. 135–64. London: Plenum Press.

Dunbar, R. I. M. (1980a). Demographic and life history variables of a population of gelada baboons (*Theropithecus gelada*). *Journal of Animal Ecology*, **49**, 485–506.

Dunbar, R. I. M. (1980b). Determinants and evolutionary consequences of dominance among female gelada baboons. *Behavioural Ecology and Sociobiology*, **7**, 253–65.

Dunbar, R. I. M. (1990). Environmental and social determinants of fecundity in primates. In *Fertility and Resources*, ed. J. Landers and V. Reynolds, pp. 5–17. Cambridge: Cambridge University Press.

Dunbar, R. I. M. (1992). Time: a hidden constraint on the behavioural ecology of baboons. *Behavioural Ecology and Sociobiology*, **31**, 35–49.

Dunbar, R. I. M. & Dunbar, P. (1988). Maternal time budgets of gelada baboons. *Animal Behaviour*, **36**, 970–80.

Estes, R. D. & Estes, R. K. (1979). The birth and survival of wildebeest calves. *Zeitschrift fur Tierpsychologie*, **50**, 45–95.

Fiesta-Bianchet, M. (1988a). Seasonal range selection in bighorn sheep: conflicts between forage quality, forage quantity, and predator avoidance. *Oecologia*, **75**, 580–6.

Fiesta-Bianchet, M. (1988b). Nursing behaviour of bighorn sheep: correlates of ewe age, parasitism, lamb age, birthdate and sex. *Animal Behaviour*, **36**, 1445–54.

Follett, B. K. (1984). The environment and reproduction. In *Reproduction in Mammals: 4 – Reproductive Fitness*, ed. C. R. Austin and R. V. Short, pp. 103–32. Cambridge: Cambridge University Press.

Frame, L. H., Malcolm, J. R., Frame, G. W. & van Lawick, H. (1979). Social organization of African wild dogs (*Lycaon pictus*) on the Serengeti plains, Tanzania, 1967–1978. *Zeitschrift fur Tierpsychologie*, **50**, 225–49.

Garber, P. A., Moya, L. & Malaga, C. (1984). A preliminary field study of the moustached tamarin monkey (*S. mystax*) in northeastern Peru: questions concerning the evolution of a communal breeding system. *Folia Primatologia*, **42**, 17–32.

Ginsberg, J. R. (1989). The ecology of female behaviour and male mating success in the Grevy's zebra. *Symposium of the Zoology Society of London*, **61**, 89–110.

Goldizen, A. W. (1987). Tamarins and marmosets: communal care of offspring. In

Primate Societies, ed. B. B. Smuts, D. L. Cheney, R. M. Seyfarth, R. W. Wrangham and T. T. Struhsaker, pp. 34–43. Chicago: University of Chicago Press.

Gomendio, M. (1989*a*). Suckling behaviour and fertility in rhesus macaques. *Journal of Zoology (London)*, **217**, 449–67.

Gomendio, M. (1989*b*). Differences in fertility and suckling patterns between primiparous and multiparous rhesus mothers (*Macaca mulatta*). *Journal of Reproduction and Fertility*, **87**, 529–42.

Goodall, J. (1983). Population dynamics during a 15 year period in one community of free-living chimpanzees in the Gombe National Park, Tanzania. *Zeitschrift fur Tierpsychologie*, **61**, 1–60.

Gordon, T. P. (1981). Reproductive behaviour in the rhesus monkey: social and endocrine variables. *American Zoologist*, **21**, 185–95.

Hadidian, J. & Bernstein, I. S. (1979). Female reproductive cycles and birth data from an Old World monkey colony. *Primates*, **20**, 429–42.

Harcourt, A. H. (1987). Dominance and fertility among female primates. *Journal of Zoology (London)*, **213**, 471–87.

Hasegawa, T. & Hiraiwa-Hasegawa, M. (1983). Opportunistic and restrictive mating among wild chimpanzees in the Mahale Mountains, Tanzania. *Journal of Ethology*, **1**, 75–85.

Hrdy, S. B. & Hrdy, D. (1976). Hierarchical relations among female hanuman langurs. *Science*, **193**, 913–15.

Hulet, C. V. & Shelton, M. (1980). Sheep and goats. In *Reproduction in Farm Animals*, ed. E. S. E. Hafez, pp. 346–57. Philadelphia, Lea & Febiger.

Jainudeen, M. R. & Hafez, E. S. E. (1980*a*). Reproductive failure in females. In *Reproduction in Farm Animals*, ed. E. S. E. Hafez, pp. 449–70. Philadelphia, Lea & Febiger.

Jainudeen, M. R. & Hafez, E. S. E. (1980*b*). Reproductive failure in males. In *Reproduction in Farm Animals*, ed. E. S. E. Hafez, pp. 471–93. Philadelphia: Lea & Febiger.

Johnson, R. L., Berman, C. M. & Malik, I. (1993). An integrative model of the lactational and environmental control of mating in female rhesus monkeys. *Animal Behaviour*, **46**, 63–78.

Kojola, I. & Eloranta, E. (1989). Influences of maternal body weight, age and parity on sex ratio in semi-domesticated reindeer (*Rangifer t. tarandus*). *Evolution*, **43**, 1331–6.

Konig, B. & Markl, H. (1987). Maternal care in house mice I. The weaning strategy as a means for parental manipulation of offspring quality. *Behavioural Ecology and Sociobiology*, **20**, 1–9.

Laurenson, M. K. (1992). Reproductive Strategies in Wild Female Cheetahs. University of Cambridge: PhD Thesis.

Lee, P. C. (1984). Early infant development and maternal care in free-ranging vervet monkeys. *Primates*, **25**, 36–47.

Lee, P. C. (1986). Early social development among African elephant calves. *National Geographic Research*, **2**, 388–401.

Lee, P. C. (1987). Nutrition, fertility and maternal investment in primates. *Journal of Zoology (London)*, **213**, 409–22.

Lee, P. C. & Moss, C. J. (1986). Early maternal investment in male and female

African elephant calves. *Behavioural Ecology and Sociobiology*, **18**, 353–61.

Lee, P. C., Majluf, P. & Gordon, I. J. (1991). Growth, weaning and maternal investment from a comparative perspective. *Journal of Zoology (London)*, **225**, 99–114.

Loudon, A. & Kay, R. (1983). Lactational constraints on a seasonally breeding mammal: the red deer. In *Physiological Strategies of Lactation*, ed. M. Peaker, R. G. Vernon and C. H. Knight, pp. 33–85. London: Academic Press.

Loy, J. (1988). Effects of supplementary feeding on maturation and fertility in primate groups. In *Ecology and Behaviour of Food-Enhanced Primate Groups*, ed. J. E. Fa and C. H. Southwick, pp. 153–66. New York: Alan R. Liss.

Lyles, A. M. & Dobson, A. P. (1988). Dynamics of provisioned and unprovisioned primate populations. In *Ecology and Behaviour of Food-Enhanced Primate Groups*, ed. J. E. Fa and C. H. Southwick, pp. 199–230. New York: Alan R. Liss.

McComb, K. (1987). Roaring by red deer stags advances the date of oestrus in hinds. *Nature, London*, **330**, 648–9.

MacDonald, D. W. & Moehlman, P. D. (1982). Co-operation, altruism and restraint in the reproduction of carnivores. In *Perspectives in Ethology, Vol 5*, ed. P. P. G. Bateson & P. H. Klopfer, pp. 433–67. New York, Plenum Press.

Malik, I. & Southwick, C. H. (1988). Feeding behaviour and activity patterns of rhesus monkeys (*Macaca mulatta*) at Tughlaqabad, India. In *Ecology and Behaviour of Food-Enhanced Primate Groups*, ed. J. E. Fa & C. H. Southwick, pp. 95–112. New York: Alan R. Liss.

Mendl, M., Zanella, A. J. & Broom, D. M. (1992). Physiological and reproductive correlates of behavioural strategies in female domestic pigs. *Animal Behaviour*, **44**, 1107–21.

Mori, A. (1979). Analysis of population changes by measurements of body weight in the Koshima troop of Japanese monkeys. *Primates*, **20**, 371–97.

Moss, C. J. & Poole, J. H. (1983). Relationships and social structure of African elephants. In *Primate Social Relationships: an Integrated Approach*, ed. R. A. Hinde, pp. 314–25. Oxford: Blackwell Scientific Press.

Nicolson, N. A. (1982). Weaning and the Development of Independence in Olive Baboons. Harvard University: PhD Thesis.

Niemeyer, C. L. & Anderson, J. R. (1983). Primate harassment of matings. *Ethology and Sociobiology*, **4**, 205–20.

Packer, C., Herbst, L., Pusey, A. E., Bygott, J. D., Hanby, J. P., Cairns, S. J. & Borgerhoff Mulder, M. (1988). Reproductive success in lions. In *Reproductive Success*, ed. T. H. Clutton-Brock, pp. 363–83. Chicago: University of Chicago Press.

Partridge, L. (1988). Lifetime reproductive success in *Drosophila*. In *Reproductive Success*, ed. T. H. Clutton-Brock, pp. 11–23. Chicago: University of Chicago Press.

Poole, J. H. (1987). Rutting behaviour in African elephants: the phenomenon of musth. *Behaviour*, **102**, 283–316.

Poole, J. H. (1989). Announcing intent: the aggressive state of musth in African elephants. *Animal Behaviour*, **37**, 140–52.

Pope, N. S., Gordon, T. P. & Wilson, M. E. (1986). Age, social rank and lactational status influence ovulatory patterns in seasonally breeding rhesus monkeys. *Biology of Reproduction*, **35**, 353–9.

Prins, H. (1987). *The Buffalo at Manyara*. Riijks University of Groningen: PhD Thesis.

Ramsay, M. A. & Stirling, I. (1988). Reproductive biology and ecology of female polar bears (*Ursus maritimis*). *Journal of Zoology (London)*, **214**, 601–34.

Riopelle, A. J. & Favret, R. (1977). Protein deprivation in primates: XIII. Growth of infants born to deprived mothers. *Human Biology*, **49**, 321–33.

Rood, J. P. (1980). Mating relations and breeding suppression in the dwarf mongoose. *Animal Behaviour*, **28**, 143–50.

Rothe, H. (1975). Some aspects of sexuality and reproduction in groups of captive marmosets (*Callithrix jacchus*). *Zeitschrift fur Tierpsychologie*, **37**, 255–73.

Rowell, T. H. (1970). Baboon menstrual cycles affected by social environment. *Journal of Reproduction and Fertility*, **21**, 133–41.

Rubenstein, D. I. (1986). Ecology and sociality in horses and zebras. In *Ecological Aspects of Social Evolution*, ed. D. I. Rubenstein & R. W. Wrangham, pp. 282–302. Princeton: Princeton University Press.

Sackett, G. P. (1990). Sires influence fetal death in pigtail macaques (*Macaca nemistrina*). *American Journal of Primatology*, **20**, 13–22.

Sackett, G. P., Holm, R. A., Davis, A. E. & Farhenbruck, C. E. (1974). Prematurity and low birth weight in pigtail macaques: incidence, prediction and effects on infant development. In *Proceedings of the Fifth Congress of the International Primatological Society*, ed. S. Kondo, M. Kawai, A. Ehara & S. Kawamura, pp. 189–206. Tokyo: Japan Science Press.

Sinclair, A. R. E. (1977). *The African Buffalo*. Chicago: University of Chicago Press.

Smith, E. F. S. (1991). The influence of nutrition and postpartum mating on weaning and subsequent play behaviour of hooded rats. *Animal Behaviour*, **41**, 513–24.

Stewart, K. J. (1988). Suckling and lactational anoestrus in wild gorillas. *Journal of Reproduction and Fertility*, **83**, 627–34.

Struhsaker, T. T. & Leyland, L. (1987). Colobines: infanticide by adult males. In *Primate Societies*, ed. B. B. Smuts, D. L. Cheney, R. M. Seyfarth, R. W. Wrangham and T. T. Struhsaker, pp. 83–97. Chicago: University of Chicago Press.

Strum, S. C. (1991). Weight and age in wild olive baboons. *American Journal of Primatology*, **29**, 219–37.

Strum, S. C. & Western, D. (1982). Variations in fecundity with age and environment in olive baboons (*Papio anubis*). *American Journal of Primatology*, **3**, 61–71.

Sugiyama, Y. & Ohsawa, H. (1982). Population dynamics of Japanese monkeys with special reference to the effects of artificial feeding. *Folia Primatologia*, **39**, 238–63.

Tanaka, I., Hayama, S.-I. & Nigi, H. (1993). Milk secretion in pregnancy among free ranging Japanese Macaques. *American Journal of Primatology*, **30**, 169–74.

Torbit, S. C., Carpenter, L. H., Swift, D. M. & Alldredge, A. W. (1985). Differential loss of fat and protein by mule deer during winter. *Journal of Wildlife Management*, **40**, 330–5.

Wasser, S. K. & Starling, A. K. (1986). Reproductive competition among female yellow baboons. In *Primate Ontogeny, Cognition and Social Behaviour*, ed. J. G. Else and P. C. Lee, P. C., pp. 343–54. Cambridge: Cambridge University Press.

Whitten, P. L. (1983). Diet and dominance among female vervet monkeys (*Cercopithecus aethiops*). *American Journal of Primatology*, 5, 139–59.

Wilson, M. E., Walker, M. L., Pope, N. S. & Gordon, T. P. (1988). Prolonged lactational infertility in adolescent rhesus monkeys. *Biology of Reproduction*, 38, 163–74.

Winkler, P., Loch, H. & Vogel, C. (1984). Life history of hanuman langurs (*Presbytis entellus*): reproductive parameters, infant mortality and troop development. *Folia Primatologia*, 43, 1–23.

Wolf, J. O. (1988). Maternal investment and sex ratio adjustment in American bison calves. *Behavioural Ecology and Sociobiology*, 23, 127–33.

4   *Evidence for interpopulation variation in normal ovarian function and consequences for hormonal contraception*

G. R. BENTLEY

Introduction

There is significant variation in the endogenous steroid hormone profiles of non-contracepting, menstruating women in different populations. Women in non-industrial societies frequently exhibit lower levels of progesterone and estrogen compared to Western subjects. This may be mediated primarily by inter-group differences in dietary composition, although we cannot rule out the possible effects of other factors such as women's current nutritional status, developmental environment, genetics, and activity patterns. There is also significant inter-population variation in the absorption, distribution, clearance and excretion rates of exogenous, synthetic contraceptive steroids. Researchers interested in such pharmacokinetic parameters have been unable to account consistently for these differences. In this chapter it is suggested that interpopulation variability in the pharmacokinetics of contraceptive steroids can be explained by similar variability in the metabolism and secretion rates of endogenous steroids.

There are a number of policy implications that arise from the hypothesis that endogenous steroid production and metabolism affect responses to exogenous or synthetic steroids. For example, inter-population differences in pharmacokinetic responses to hormonal contraceptives cause variable side effects such as intermenstrual bleeding which can severely impact contraceptive continuance rates. Understanding the causes of interpopulation responses to exogenous hormones may therefore enable more suitable doses for specific populations to be recommended than those currently used and, hence, improve policy effectiveness. In addition, the doses thought to be required for contraceptive efficacy in different areas of the world are often based on studies of Western women who may be poor models for other populations. In fact, these doses may be inappropriate in

46

many cases. By ignoring normal biological variation in reproductive hormonal systems and its causes, we may therefore be inadvertently elevating the risk of physiological complications for women in many countries who take standard contraceptive doses.

The evidence for significant differences between populations in levels of reproductive steroids will first be discussed, followed by a discussion of the potential causes for these differences. Focus is primarily on the possibility that variation in dietary composition may affect the metabolic processes of endogenous steroids, resulting in the interpopulation variability recorded. The significant differences that have been recorded between populations in their pharmacokinetic responses to exogenous contraceptive steroids are summarized, followed by a discussion of the possible reasons for these differences. A third section concerns the differential side-effects experienced by steroid contraceptive users. A major issue here is whether such side-effects could be predicted by having prior knowledge of women's natural steroid levels and metabolic patterns, resulting in the recommendation of specific doses for different populations. The chapter concludes with comments on the policy implications that arise from this possibility.

Interpopulation variation in reproductive steroids

There are significant interpopulation differences in levels of reproductive steroids measured in saliva, plasma, urine and feces. Research in this area has been undertaken by biological anthropologists (primarily reproductive ecologists) and by clinicians. The former tend to collect longitudinal data from small populations under difficult field conditions. They have concentrated more on the analysis of salivary progesterone for which reliable assays exist, since salivary estradiol assays remain problematic to date (Read, 1993). In contrast, clinicians have focused more on short-term surveys of much larger groups of women, collecting urinary, fecal, plasma and serum samples of estrogens.

Data collected by reproductive ecologist Peter Ellison and his colleagues (Ellison, Peacock & Lager, 1986, 1989; Ellison *et al.*, 1993, Jasienska & Ellison, 1993; Panter-Brick, Lotstein & Ellison, 1993; Vitzthum, Ellison & Sukalich, 1994) for populations in Zaire, North America, Poland, Nepal and Bolivia illustrate clear differences in mean levels of salivary progesterone (Fig. 4.1). Women in the more industrialized countries with better nutrition, such as American and Polish women, show significantly higher levels of salivary progesterone compared to the more traditional, rural populations like the Lese who are slash-and-burn horticulturalists in the

48   G. R. Bentley

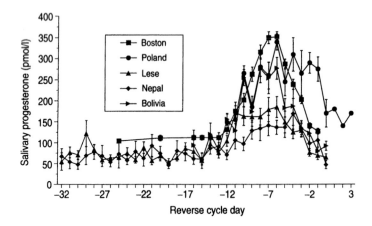

Fig. 4.1. Inter-population differences in salivary progesterone profiles for
non-contracepting women. (Data provided by Peter T. Ellison, Grazyna
Jasienska, Catherine Panter-Brick and Virginia Vitzthum.)

Ituri Forest of northeastern Zaire, and the Tamang who are agro-
pastoralists in the Himalayan foothills of northwestern Nepal. Mean
average mid-luteal progesterone (calculated as cycle day −5 to −9) is
approximately 372 pmol/l for healthy, well-nourished American women
from Boston, 252 pmol/l for agricultural women from Bolivia (Vitzthum,
Ellison & Sukalich, 1994), 201 pmol/l for the Lese (Ellison, Peacock &
Lager, 1989), 185 pmol/l for the Tamang (Panter-Brick, Lotstein & Ellison,
1993), and pmol/l from pilot data for Polish agricultural women (Jasienska
& Ellison, 1993). An important feature for these studies is that they were,
undertaken in the same laboratory, thus reducing the possibility of assay
variation and methodological noise that can create significant problems for
comparative studies. A limited number of plasma and salivary studies of
reproductive steroids in other non-industrial populations have also
confirmed low profiles of progesterone and estradiol (e.g. Seaton &
Riad-Fahmy, 1980; Van der Walt, Wilmsen & Jenkins, 1978; Van der Walt
et al., 1977; Worthman et al., 1993).

These interpopulation contrasts in reproductive steroids have been
related to chronic differences in energy balance, primarily translated
through nutrition and activity patterns (Bailey et al., 1992; Ellison,
Peacock & Lager, 1989, 1993; Jasienska & Ellison, 1993; Panter-Brick,
Lotstein & Ellison, 1993). This supposition is supported by clinical findings
of a reduction in reproductive steroid levels among dieting and exercising
Western women (e.g. Cumming, Wheeler & Harber, 1994; Rosetta, 1993;
Warren, 1983). However, it is not clear that lower levels of salivary steroids

in the non-Western groups reflect chronic adaptation to poor nutritional conditions that would be ameliorated if these women could enjoy a better diet. They might instead reflect a different adult setting of ovarian function mediated by individual responses to environmental conditions experienced during critical periods of growth and development, such that women may become 'adapted to' specific circulating levels of reproductive hormones (Ellison, 1990, 1994; Vitzthum, 1990). Further studies must be undertaken to shed more light on this topic. Some deductions can be made, however, from more clinically oriented studies of interpopulation variation in steroid levels, and particularly by migrant studies discussed further below.

A different explanation in the clinical literature for why significant interpopulation variation in reproductive steroid levels may exist concerns the relationship between dietary composition and estrogen levels. Most of this comparative clinical work has arisen from research into causal factors behind interpopulation variation in cancer rates rather than from an intrinsic interest in the link between steroid levels and ovarian function. The concentration on estrogens results from the well-established link between this steroid and breast cancer risks (Bernstein & Ross, 1993; Cole & MacMahon, 1969; Longcope, 1990). There has also been a focus on Asian populations where rates for all cancers are significantly lower than Western groups. In comparison to Western controls, premenopausal Asian women from Japan, China and Korea have lower serum and urinary levels of oestradiol and oestrone, while oestriol levels are sometimes, but not always, lower (Briggs & Briggs, 1972; Goldin *et al.*, 1986; Key *et al.*, 1990; MacMahon *et al.*, 1971; Shimuzi *et al.*, 1990).

These studies relate such interpopulation differences to variation in dietary composition, and particularly the amount of fat and fibre in the diet. Traditional Asian diets tend to be lower in total and saturated fats than Western diets, and are more comparable to vegetarian diets in the proportions of macronutrients (Goldin *et al.*, 1986; Hill *et al.*, 1977; MacMahon *et al.*, 1974). In support of this association, Western vegetarian women also have lower levels of plasma and urinary oestradiol and oestrone, and higher total faecal estrogens compared to omnivorous controls (Armstrong *et al.*, 1981; Goldin *et al.*, 1982; Shultz & Leklem, 1983). Asian migrants, mostly living in Hawaii, who have been studied both for cancer risks and oestrogen levels show intermediate values compared to their Asian counterparts who still follow a traditional diet within their country of origin (Dickinson *et al.*, 1974; Trichopoulos *et al.*, 1984).

What are the potential causes of variation in dietary composition on steroid metabolism, bearing in mind that both endogenous and exogenous steroid metabolism will be affected by these differences? First, both oestrogens and, to a lesser extent progestins, go through enterohepatic circulation where the liver secretes steroids into the bile for transportation to the intestines. Conjugation to glucuronides can also occur in the liver causing fecal excretion of the conjugate, or intestinal deconjugation can liberate the steroid for reabsorption. Enterohepatic cycling, which can be repeated several times, results in a reduced and variable range of steroid bioavailability (Adlercreutz & Martin, 1980; Gibaldi, 1991; Newburger & Goldzieher, 1985). Both dietary composition and individual differences in these metabolic processes can affect enterohepatic cycling and systemic levels of steroids.

The intestines are evidently an important site of steroid metabolism and conversion (Adlercreutz *et al.*, 1976, 1979). For example, the composition of intestinal flora is known to affect steroid metabolism, intestinal absorption, and the excretion rate of faecal steroids, and can be changed by dietary factors. For example, a vegetarian diet can lead to diminished recycling of oestrogen and the excretion of a larger amount of biliary oestrogens (Goldin *et al.*, 1981, 1982). Armstrong *et al.* (1981) have also suggested that vegetarian diets might increase the 2-hydroxylation of oestradiol, and perhaps decrease 16α-hydroxylation, thus influencing the amount of circulating oestrogens. Such metabolic conversions are important for oestrogen availability since increased production of 2-hydroxyestrone reduces the peripheral action of oestradiol, while an increase in 16α-hydroxyestrone increases its potency (Fishman & Martucci, 1980).

Experimental studies undertaken to evaluate metabolic changes in reproductive steroid levels using dietary manipulations also conclude that dietary composition is an important determinant of exogenous and endogenous steroid metabolism. Musey *et al.* (1987) have studied the effects of different diets on the oxidative metabolism of 17β estradiol in chimpanzees, primates which metabolize steroids in similar fashion to humans. A high fat diet (65%) significantly decreased the 2-hydroxylation, and increased the 16α-hydroxylation of estradiol. Longcope *et al.* (1987) performed a similar experiment using human subjects, where six healthy women aged 21–32 were fed a high-fat diet for two months (40% of total calories from fat) followed by a low-fat diet for two months (25% of total calories). The low fat diet reduced the 16α-hydroxylation of oestradiol. Other studies have shown a reduction in serum total oestrogens as well as oestrone and oestradiol when women have been following a low-fat diet, suggesting similar metabolic processes (Rose *et al.*, 1987).

The proportion of dietary fat ingested may not be the only factor affecting steroid metabolism. For example, Kappas *et al.* (1983) and

Anderson *et al.* (1984) studied the effect of changing dietary protein-to-carbohydrate ratios on the metabolism of testosterone and oestradiol among eight males. These subjects ate a high-carbohydrate diet for two weeks, followed by a high protein diet for a second two-week period with an interval in between. The high protein diet significantly increased the 2-hydroxylation of oestradiol. Fishman, Boyar and Hellman (1975) reported earlier that there is an inverse correlation between body weight and 2-hydroxylation of estradiol, although this may be related to fat intake. Williams and Goldzieher (1980) have also shown that there is regional variability in the hepatic metabolism of ethinyl oestrogens among human populations which has a bearing on interpopulation differences in steroid levels. They found greater conjugation of ethinyl oestrogens to monoglucuronides among American women, greater conjugation to diglucuronides among Nigerian women, and an equal proportion between these conjugates among women from Sri Lanka. The degree of oxidative metabolism showed similar variation. All these *in vivo* studies summarized above confirm that dietary composition is a critical factor explaining interpopulation variation in steroid profiles.

There are some indirect supporting data for the role of nutritional composition affecting steroid levels for women in the populations who have been compared for their salivary progesterone levels by Ellison and his colleagues. Lese women from the Ituri Forest of Zaire, for example, eat a diet that is low in fat intake (11%), with a high protein-to-carbohydrate ratio of approximately 5:1 (Bentley *et al.*, 1992). The Lese diet is also mostly vegetarian, since women consume a relatively small amount of either fish or meat, averaging 12% of the total diet. Meat intake for individual Lese women, however, is also highly variable due to individual food taboos against certain forest animals, many of which are also prohibited for consumption during the reproductive periods of a women's lifetime (Aunger, 1992). Tamang women in Nepal, ingest a diet that has an even lower fat content (4%) than the Lese, with a protein-to-carbohydrate ratio of almost 1:8 (Koppert, 1988). The body mass index is also slightly lower for Tamang women (20.83) compared to Lese women (21.87). From the salivary steroid comparisons made by Ellison among different populations, both the Lese and the Tamang also have the lowest salivary progesterone profiles.

The metabolism of endogenous reproductive steroids has not yet been examined in field settings among the more traditional, subsistence-based populations like the Lese and Tamang. Admittedly, this kind of research would be extremely complicated to conduct in the isolated areas where these groups live. But such studies could shed light on the relationship between production, absorption and clearance of endogenous reproductive steroids, how metabolic rates might vary between different populations, and how important these processes are in explaining variation in reproductive steroid profiles. It is possible that chronic differences in progesterone levels merely reflect differences in nutrient composition in conjunction with caloric stress. Variations in other indices of ovarian function which are also dependent on circulating hormones, such as length of the follicular and luteal phases, menstrual cycle length and days of menstrual flow may also provide clues to interpopulation differences in steroid metabolism. The implications of differential steroid metabolism and dietary composition for the exogenous steroid contraceptives will be discussed further below.

Steroid contraceptives and their pharmacokinetic parameters

Researchers involved in studies of steroid contraceptives have reported wide inter-population variation among women in their pharmacokinetic responses to synthetic hormonal compounds even when using the same kind and dose of hormonal contraceptive (e.g. Fotherby, 1983; Fotherby et al., 1979, 1981; Garza-Flores, Hall & Perez-Palacios, 1991; Goldzieher, 1989; Goldzieher, Dozier & de la Pena, 1980; Kuhl, 1990; Sang et al., 1981; Stadel et al., 1980; Yong-en et al., 1987). The most common pharmacokinetic properties that are examined in studies of steroid contraceptives include, but are not limited to, maximal serum concentration (C_{max}), time to maximum serum concentration (T_{max}), the area under the serum concentration versus time curve (AUC), and the serum elimination half-life ($T_{1/2el}$).

Steroid contraceptives currently in use include oral compounds (combined oestrogen and progestogen, or progestin-only pills), long-acting progestogen injectables such as depot-medroxyprogesterone acetate (DMPA; Depo-Provera®) and norethisterone enanthate (NET-EN), steroid implants releasing the progestogen levonorgestrel (e.g., Norplant®), and steroid-releasing vaginal rings. There has been a significant decline in recommended doses in the last 30 years, particularly in the oral preparations, as researchers discovered that significant side effects could be reduced without compromising contraceptive efficacy (Gerstman et al., 1991). In fact, some researchers acknowledge that different countries may require specific consideration of the safety issues involved in contraceptive research, and that health policy decisions made in the developed world cannot necessarily be transported uncritically to less developed areas (Rivera, 1993).

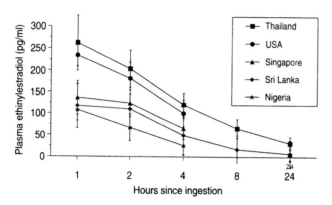

Fig. 4.2. Plasma levels (corrected for body surface) of unconjugated ethinylestradiol in five populations of women after ingesting 50 ml. (Data taken from Goldzieher *et al.*, 1980.)

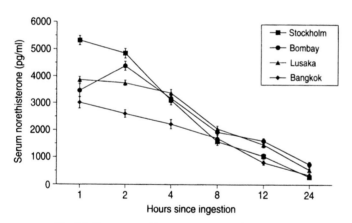

Fig. 4.3. Serum norethisterone concentrations in four populations of women after ingesting 1 mg. (Data taken from Fotherby *et al.*, 1979.)

There are a number of reports of interpopulation variation in pharmacokinetic properties. For example, ethinyl oestradiol was administered to women in North America, Nigeria, Singapore, Sri Lanka and Thailand, (Goldzieher, Dozier & de la Pena, 1980). There were statistically significant intergroup differences in plasma levels of this steroid measured at various intervals after dosage, even after correcting for body size (Fig. 4.2). Fotherby *et al.* (1981) conducted a similar study and also found wide intergroup variation in plasma levels of ethinyl oestradiol. Likewise, the metabolic clearance rates of orally administered norethisterone differed significantly between populations (Fig. 4.3; Fotherby *et al.*, 1979). Other pharmacokinetic parameters, such as the absorption and elimination half-lives and the area under the concentration time curve show similar variation (Fotherby, Hall & Pereg-Palacios, 1979; Garza-Flores, Dozier & de la Pena, 1991; Goldzieher *et al.*, 1980; Yong En, 1987; Stadel *et al.*, 1980).

Researchers have also reported large *interindividual* differences in pharmacokinetic responses to exogenous steroids (e.g. Bergink *et al.*, 1990; Carol *et al.*, 1991; Fotherby *et al.*, 1979, 1981; Goldzieher, Dozier & de la Pena, 1980; Prasad *et al.*, 1981; Stadel *et al.*, 1980). For example, the coefficient of variation for AUC_{0-24} was 41% among women taking an oral dose of ethinyl oestradiol in two 35 μg pills, while the range was -66% to $+71\%$ (Goldzieher & Brody, 1990). Fotherby *et al.* (1981) found a range of 2.5–30 hours in $T_{1/2}$el of a 50 μg oral dose of ethinyl oestradiol among individual women in their study of 15 different populations. Sang *et al.* (1981) also found a large range of 7.5–22.5 hours in the $T_{1/2}$el of ethinyl oestradiol in their study of nine subjects from a single population. Similar coefficients of variation and ranges have been found as well as large standard deviations in pharmacokinetic parameters for other contraceptive steroids (Bergink *et al.*, 1990). The existence of large interindividual variations does not appear, however, to influence either the existence or the interpretation of interpopulation differences (Goldzieher & Brody, 1990), although these might result from many of the same causes that affect the latter.

Variability in pharmacokinetic responses can result from many factors, such as the site and method of drug administration, degrees of steroid binding to plasma proteins like sex hormone binding globulin, time since food ingestion, genetic differences, differences in muscle-to-fat ratios that might affect storage and metabolism of steroids, and methodological differences in evaluating pharmacokinetic parameters (Bergink *et al.*, 1990; Goldzieher, Dozier & de la Pena, 1980; Goldzieher & Brody, 1990; Kaufman, Thiery & Vermeulen, 1981; Newburger & Goldzieher, 1985). Differences in body weight between women may also be important. Prasad *et al.* (1981) found significant positive correlations between body weight and post-peak average plasma NET–EN concentration, and mid-arm circumference and average plasma NET–EN concentration. They attribute this relationship to the fact that thinner women with a lean body and muscle mass exhaust the contraceptive depot of NET–EN earlier than fatter subjects.

By far the largest effect, however, may result from differences in individual metabolism of exogenous steroids, and the effect of dietary composition on these metabolic processes (Gibaldi, 1991; Newburger & Goldzieher, 1985). For example, ethinyl oestradiol, desogestrel, norethisterone (but not levenorgestrel or gestodene) are subject to first pass effects, or presystemic metabolism, through the enterohepatic circulation (Goldin *et al.*, 1981; Goldzieher, 1989; Shenfield & Griffin, 1991; Stanczyk *et al.*, 1983, 1990), processes analogous to those discussed earlier involving the endogenous steroids. (This first-pass effect can, in fact, be avoided by parenteral administration of steroid contraceptives (Kuhl, 1990).) Presystemic metabolism involves transportation of exogenous steroids through the gastrointestinal tract, the gut wall and finally the liver before they become available for delivery by the blood system. Steroid metabolism can occur during passage through the gastrointestinal tract, during absorption in the intestinal walls and, finally, in the liver itself, meaning that a certain proportion of the steroids is unavailable for systemic circulation. Again, dietary factors can affect gastrointestinal absorption of exogenous steroids, underscoring the importance of variation in nutritional composition between populations.

Steroid contraceptives and ovarian function

Differences between steroid compounds and their pharmacokinetic properties also affect the ovarian function of women using them. For example, many women using low-dose progestogens (including orals, injectables and implants) experience normal ovulatory cycles, rather than ovulatory suppression or ovulatory cycles with luteal insufficiency (Brache *et al.*, 1990; Landgren & Diczfalusy, 1980; Paulsen *et al.*, 1974; Prasad *et al.*, 1981). The contraceptive effects of the progestogens thus appear to act primarily through changes in normal function of the vaginal mucosa, endometrium, Fallopian tubes and corpus luteum (Fotherby, Svendsen & Foss, 1968; Johannisson, Landgren & Diczfalusy, 1982; Wright, Fotherby & Fairweather, 1970). Cycle characteristics of non-contracepting women could, therefore, have a predictive impact on the evaluation of women's potential metabolic responses to steroid contraceptives, and hence could be useful for assessing dose recommendations. Dickey (1979) suggests that menstrual characteristics are also helpful in determining recommendations for oral preparations.

Clinical studies of contraceptive users, however, do not usually report the typical profiles of endogenous reproductive steroids or their metabolism in women prior to their use of synthetic compounds. But, there are a few studies that support the link between normal cycle characteristics and a woman's response to steroid contraceptives. In fact, Landgren and Diczfalusy (1980) describe four distinct categories of ovarian response to 300 μg of orally administered norethisterone among a sample of 43 women.

Table 4.1. *Indices of ovarian function in Swedish women in a control and treatment cycle with norethisterone (300 μg* daily)

Indices of ovarian function	Group A (n=7)		Group B (n=17)	
	Control	Treatment	Control	Treatment
Mean estradiol (pmol/l) for days 1–6 of cycle	240	251	244	282
Luteal maximum of oestradiol (pmol/l)	690	n.d.	770	680
Luteal maximum of progesterone (nmol/l)	39.2*	<1.4	60.4	45.7
Luteal mean progesterone (nmol/l)	18.8*	<1.4	28.8	22.6
Length of follicular phase (days)	18.6*	n.d.	12.6	15.1
Length of luteal phase (days)	11.7*	n.d.	15.8	13.5
Cycle length (days)	32.2	35	27.5	28.8
Average number of bleeding-free days in cycle	26.7	17.5	22.5	17.4
Mean plasma levels of norethisterone (pmol/l) 19–24 hours after ingestion		773		873

Group A includes women with lower indices of ovulatory function in the control cycles, while Group B includes women with higher indices. Data taken from B.M. Landgren and E. Diczfalusy, 1980. Hormonal effects of the 300 μg norethisterone (NET) minipill. *Contraception* **21**(1), 87–113.
* Indices significant differences (p < 0.05) between control groups.

Category A included women in whom both luteal and follicular activity were completely suppressed, Category B exhibited cyclic follicular but completely suppressed luteal activity, Category C included women with cyclic follicular and diminished luteal activity, while Category D included women with both normal luteal and follicular cycles. In this study, there was an evident relationship between ovarian function *prior* to steroid contraceptive use and ovarian function during ingestion of oral doses. Women with lower levels of oestradiol and progesterone, long follicular and short luteal phases, and fewer days of menstrual bleeding during control cycles tended to exhibit anovulatory cycles during treatment with norethisterone. In contrast, women with higher levels of oestradiol and progesterone, shorter follicular phases, adequate luteal phases and longer periods of menstrual bleeding tended to have ovulatory cycles during treatment (Table 4.1). Many of these differences between the two groups were significant.

It is also possible to deduce some supporting information regarding different populations from other published data. For example, Fotherby, Koetsawang and Mathrubutham (1980) examined the pharmacokinetic and pharmacodynamic responses to DMPA and NET–EN injection among eight Swedish and eight Indian women. Plasma levels of norethisterone were detectable for significantly longer in Swedish compared to Indian women, and Indian women returned to ovulation significantly sooner than the Swedes. Examination of the endogenous steroid profiles of these two populations shows that plasma peak progesterone values for Indian women in their control cycles are much lower (7 ng/ml) and peak oestradiol levels higher (315 pg/ml) than progesterone (15 ng/ml) and oestradiol (250 pg/ml) levels in the Swedish subjects. All of these data indicate a link between normal ovarian cycle characteristics of women and their pharmacokinetic responses to steroid contraceptives. They also support the idea that predictions about women's responses to contraceptive steroids could be made from observing normal cycle characteristics.

Contraceptive side-effects and policy implications

A major clinical goal has been the reduction, if not elimination, of troublesome side effects as a means of improving rates of steroid contraceptive use, although the primary concern in many developing countries lies in pregnancy prevention and population control through contraceptive efficacy. Yet, if predictions could be made of potential side effects in different populations through the evaluation of endogenous steroid levels prior to contraceptive use, followed by better recommendations for efficacious and safe contraceptive steroid doses, then family planning programs in different regions of the world might meet with greater success than is currently the case. There are a wide variety of major and minor side effects reported by contracepting women in different populations, even when the same type of contraceptive compound is used. These side effects include thromboembolic disorders, altered carbohydrate and lipid metabolism, nausea, dizziness, weight gain, bloating, breast tenderness, irritability, depression, lassitude, and loss of libido (e.g. Adams Hillard, 1989; Guillebaud, 1986; WHO, 1978). By far the most intractable side effect, mainly associated with the low-dose progestogens, is intermenstrual bleeding, which leads to high rates of contraceptive discontinuance (Diaz *et al.*, 1982; Faundes, Sivin & Stern, 1978; WHO, 1978; 1982*a,b*). This also varies in occurrence between populations. Bye (1975), for example, noted a 5.8% incidence of intermenstrual bleeding among British women taking a combined oral contraceptive consisting of 150 μg D-norgestrel and 30 μg ethinyl estradiol, while Apelo and Veloso (1975) reported a much lower incidence of 2.0% among Philippino women from

Manila using exactly the same contraceptive preparation. Similarly, Bergstein (1975) compiled a comparative list of studies reporting breakthrough bleeding and other side effects which show equally wide interpopulation variation.

There are indications that body weight may be an important factor in determining the incidence of side effects (e.g. Talwar & Berger, 1977). Gray (1980) found that women over 62 kg using NET–EN injectables suffered from more intermenstrual bleeding than women weighing less than 47 kg. This was not the case for women using DMPA. The relationship between body weight and the occurrence of side effects is evidently not a simple one, judging by differences in reports. Evidently, the type and dose of steroid contraceptive used is important, and there may be significant interstudy variation in the accuracy of reports based on cultural perceptions and methodological differences. In addition, older contraceptive studies rarely report the body mass index as a variable which would be a much better comparative index of body composition. Nor are there any studies examining differences in dietary composition that might be an important factor in determining interpopulation differences, although Fotherby et al. (1981) have suggested that there might be more breakthrough bleeding in women with a lower bioavailability and a shorter half-life of elimination of ethinyl oestradiol. This clearly links variation in the experience of side effects to individual metabolic responses to steroid contraceptives.

Certainly, the desire for effective contraceptive control has sometimes led to the use of higher doses of steroids in countries other than the West with possible resulting health risks (e.g. Lindsay, 1991). In fact, clinical studies of hormonal contraceptives have shown that even standard doses sometimes exceed the expected contraceptive effectiveness in different populations. For example, Bassol et al. (1984) examined ovarian function in Mexican women taking different doses of DMPA. They found that the standard dose of 150 mg suppressed ovulation for twice as long (180 days) as its advertised properties (90 days). Similarly, in a group of Thai women taking similar doses of DMPA, cycles were suppressed for 136 days (Fotherby, Koetsawang & Mathrubutham, 1980). These studies recommended lowering the doses in these countries. Current doses of synthetic steroids, therefore, may be subjecting women in some countries to unacceptably high side-effects and health risks. Similarly, Stadel et al. (1980) have recommended that contraceptive steroid doses should be adjusted against specific individual reactions, although it is not quite clear how feasible this would be on a financial level.

Conclusions

It has been suggested in this paper that interpopulation variation in endogenous steroid levels and ovarian function may explain similar interpopulation variation in the pharmacokinetic response to exogenous contraceptive steroids. Studies on the metabolism of synthetic and endogenous steroids have already shown a close similarity between the two processes, particularly for ethinyl estradiol (Fotherby, 1975). Researchers undertaking clinical contraceptive trials have controlled for the effects of environment, age, and ethnic origin on pharmacokinetic variability, by examining covariates such as weight, height, race and so forth. But, so far none of these variables have been able to account consistently for the significant interpopulation differences observed (Fotherby *et al.*, 1979; Goldzieher, 1989; Prasad *et al.*, 1981; Stadel *et al.*, 1980). Furthermore, it is suggested that interpopulation differences in dietary composition may be affecting the metabolism of both endogenous and exogenous reproductive steroids, as well as variation in contraceptive side effects. This hypothesis deserves further development and scrutiny.

There are several policy implications that arise from these connections. Evaluation of the metabolism of endogenous reproductive steroids and ovarian function in different populations may help to predict responses by these groups to exogenous contraceptive steroids. This could in turn lead to recommendations for specific and effective contraceptive doses better suited to different populations, to a reduction in potential health risks, unpleasant side effects and dropout rates. What remains now is to proceed with ways of testing further the hypotheses presented here so that any practical applications for contraceptive research can be developed.

ACKNOWLEDGEMENTS

An earlier version of this paper was presented in poster-format at a New York Academy of Sciences conference on Human Reproductive Ecology in May 1993 (Bentley, 1994), and as a spoken paper at the General Conference of the International Union for the Scientific Study of Population in Montreal the same year. I am grateful to my colleagues Peter Ellison, Grażyna Jasieńska, Catherine Panter-Brick and Virginia Vitzthum for permission to use unpublished data. Robert Aunger provided help in preparing the figures, and in criticizing different stages of the manuscript.

REFERENCES

Adams Hillard, P. J. (1989). The patient's reaction to side effects of oral contraceptives. *American Journal of Obstetrics and Gynecology*, 161, 1412–15.

Adlercreutz, H. & Martin, F. (1980). Biliary excretion and intestinal metabolism of progesterone and estrogen in man. *Journal of Steroid Biochemistry*, 13, 231–44.

Adlercreutz, H., Martin, F., Pulkkinen, M., Dencker, H., Rimér, U., Sjöberg, N. O. & Tikkanen, M. J. (1976). Intestinal metabolism of estrogens. *Journal of Endocrinology and Metabolism*, 43, 497–505.

Adlercreutz, H., Martin, F., Järvenpää, P. & Fotsis, T. (1979). Steroid absorption and enterohepatic recycling. *Contraception*, 20, 201–23.

Anderson, K. E., Kappas, A., Conney, A. H., Bradlow, H. L. & Fishman, J. (1984). The influence of dietary protein and carbohydrate on the principal oxidative biotransformations of estradiol in normal subjects. *Journal of Endocrinology and Metabolism*, 59, 103–7.

Apelo, R. & Veloso, I. (1975). Clinical experience with ethinyl estradiol and d-norgestrel as an oral contraceptive. *Fertility and Sterility*, 26, 283–8.

Armstrong, B. K., Brown, J. B., Clarke, H. T., Hahnel, R., Masarei, J. R. & Ratajczak, T. (1981). Diet and reproductive hormones: a study of vegetarian and nonvegetarian postmenopausal women. *Journal of the National Cancer Institute*, 67, 761–7.

Aunger, R. (1992). The nutritional consequences of rejecting food in the Ituri Forest of Zaire. *Human Ecology*, 20, 263–91.

Bailey, R. C., Jenike, M. R., Ellison, P., Bentley, G. R., Harrigan, A. M. & Peacock, N. R. (1992). The ecology of birth seasonality among agriculturalists in central Africa. *Journal of Biosocial Science*, 24, 393–412.

Bassol, S., Garza-Flores, J., Cravioto, M. C., Diaz-Sanchez, V., Fotherby, K., Lichtenberg, R. & Perez-Palacios, G. (1984). Ovarian function following a single administration of depomedroxyprogesterone acetate (DMPA) at different doses. *Fertility and Sterility*, 42, 216–22.

Bentley, G. R. (1994). Ranging hormones: do hormonal contraceptives ignore human biological variation and evolution? In *Human Reproductive Ecology: Interactions of Environment, Fertility and Behaviour*, ed. K. L. Campbell and J. W. Wood. New York: New York Academy of Sciences. *Annals of the New York Academy of Sciences*, 709, 201–3.

Bentley, G. R., Harrigan, A. M., Aunger, R. V. & Ellison, P. T. (1992). Nutritional effects on Lese ovarian function. Paper presented at an Invited Session: *Ecological Perspectives on Human Fertility*. 91st Annual Meeting of the American Anthropological Association, San Francisco, California.

Bergink, W., Assendorp, R., Kloosterboer, L., van Lier, W., Voortman, G. & Qvist, I. (1990). Serum pharmacokinetics of orally administered desogestrel and binding of contraceptive progestogens to sex hormone-binding globulin. *American Journal of Obstetrics and Gynecology*, 163, 2132–7.

Bergstein, N. A. M. (1975). Clinical efficacy, acceptability and metabolic effects of new low dose combined oral contraceptives. *Acta Obstetrica et Gynecologica Scandinavia*, Supplement 54, 51–9.

Bernstein, L. & Ross, R. K. (1993). Endogenous hormones and breast cancer risk. *Epidemiological Reviews*, 15, 48–65.

Brache, V., Alvarez-Sanchez, F., Faundes, A., Tejada, A. & Cochon, L. (1990). Ovarian endocrine function through five years of continuous treatment with Norplant subdermal contraceptive implants. *Contraception*, **41**, 169–77.

Briggs, M. H. & Briggs, M. (1972). Steroid hormone concentrations in blood plasma from residents of Zambia, belonging to different ethnical groups. *Acta Endocrinologica*, **70**, 619–24.

Bye, P. G. T. (1975). Analysis of a multicentre trial of a new low-dose contraceptive in Great Britain. *Acta Obstetrica et Gynecologica Scandinavia*, Supplement **54**, 54–66.

Carol, W., Klinger, G., Jäger, R., Kasch, R. & Michels, W. (1991). Untersuchungen zur inter-und intraindividuellen variabilität pharmakokinetischer parameter für kontrazeptive steroide. *Zentralblatt für Gynäkologie*, **113**, 783–7.

Cole, P. & MacMahon, B. (1969). Oestrogen fractions during early reproductive life in the aetiology of breast cancer. *The Lancet*, March 22, 604–6.

Cumming, D. C., Wheeler, G. D. & Harber, V. J. (1994). Physical activity, nutrition and reproduction. In *Human Reproductive Ecology: Interactions of Environment, Fertility, and Behavior*, ed. K. L. Campbell and J. W. Wood. New York: New York Academy of Sciences. *Annals of the New York Academy of Sciences*, **709**, 55–76.

Diaz, S., Pavez, M., Miranda, P., Robertson, D. N., Sivin, I. & Croxatto, H. B. (1982). A five-year clinical trial of levonorgestrel silastic implants (Norplant). *Contraception*, **25**, 447–57.

Dickey, R. P. (1979). Initial pill selection and managing the contraceptive pill patient. *International Journal of Gynaecology and Obstetrics*, **16**, 547–55.

Dickinson, L., MacMahon, B., Cole, P. & Brown, J. B. (1974). Estrogen profiles of Oriental and Caucasian women in Hawaii. *The New England Journal of Medicine*, **291**, 1211–13.

Ellison, P. T., Peacock, N. R. & Lager, C. (1989). Ecology and ovarian function among Lese women of the Ituri Forest, Zaire. *American Journal of Physical Anthropology*, **78**, 519–26.

Ellison, P. T., Peacock, N. R. & Lager, C. (1986). Salivary progesterone and luteal function in two low-fertility populations of northeast Zaire. *Human Biology*, **58**, 473–83.

Ellison, P. T. (1994). Developmental influences on adult reproductive function. *American Journal of Physical Anthropology*, Supplement **18**, 85.

Ellison, P. T. (1990). Human ovarian function and reproductive ecology: new hypotheses. *American Anthropologist*, **92**, 933–52.

Ellison, P., Lipson, S. F., O'Rourke, M. T., Bentley, G. R., Harrigan, A. M., Panter-Brick, C. & Vitzthum, V. J. (1993). Population variation in ovarian function. *The Lancet*, **342**, 433–4.

Faundes, A., Sivin, I. & Stern, J. (1978). Long acting contraceptive implants: an analysis of menstrual bleeding patterns. *Contraception*, **18**, 355–65.

Fishman, J., Boyar, R. M. & Hellman, L. (1975). Influence of body weight on estradiol metabolism in young women. *Journal of Clinical Endocrinology and Metabolism*, **41**, 989–91.

Fishman, J. & Martucci, C. (1980). Biological properties of 16α-hydroxyestrone: implications in estrogen physiology and pathophysiology. *Journal of Clinical Endocrinology and Metabolism*, **51**, 611–15.

Fotherby, K. (1975). Metabolism of synthetic steroids by animals and man. *Acta Endocrinologica*, Supplement **185**, 119–47.

Fotherby, K. (1983). Variability of pharmacokinetic parameters for contraceptive steroids. *Journal of Steroid Biochemistry*, **19**, 817–20.

Fotherby, K., Akpoviroro, J., Abdel-Rahman, H. A., Toppozada, H. K., Coutinho, E. M. *et al.* (1981). Pharmacokinetics of ethynyloestradiol in women from different populations. *Contraception*, **23**, 487–97.

Fotherby, K., Koetsawang, S. & Mathrubutham, M. (1980). Pharmacokinetic study of different doses of Depo Provera. *Contraception*, **22**, 527–38.

Fotherby, K., Saxena, B. N., Shrimanker, K., Hingorani, V., Takker, D., Diczfalusy, E. & Landgren, B. (1980). A preliminary pharmacokinetic and pharmacodynamic evaluation of depot-medroxyprogesterone acetate and norethisterone oenanthate. *Fertility and Sterility*, **34**, 131–9.

Fotherby, K., Shrimanker, K., Abdel-Rahman, H. A., Toppozada, H. K., de Souza, J. C., Coutinho, E. M. *et al.* (1979). Rate of metabolism of norethisterone in women from different populations. *Contraception*, **19**, 39–45.

Fotherby, K., Svendsen, E. K. & Foss, G. L. (1968). Ovarian function in women receiving low doses of a synthetic progestin, norgestrel. *Journal of Reproduction and Fertility*, Supplement **5**, 155–66.

Garza-Flores, J., Hall, P. E. & Perez-Palacios, G. (1991). Long-acting hormonal contraceptives for women. *Journal of Steroid Biochemistry and Molecular Biology*, **40**, 697–704.

Gerstman, B. B., Gross, T. P., Kennedy, D. L., Bennett, R. C., Tomita, D. K. & Stadel, B. V. (1991). Trends in the content and use of oral contraceptives in the United States, 1964–88. *American Journal of Public Health*, **81**, 90–6.

Gibaldi, M. (1991). *Biopharmaceutics and Clinical Pharmacokinetics*. Fourth Edition. Philadelphia, PA: Lea and Febiger.

Goldin, B. R., Adlercreutz, H., Dwyer, J. T., Swenson, L., Warram, J. H. & Gorbach, S. L. (1981). Effect of diet on excretion of estrogens in pre- and post-menopausal women. *Cancer Research*, **41**, 3771–3.

Goldin, B. R., Adlercreutz, H., Gorbach, S. L., Warram, J. H., Dwyer, J. T., Swenson, L. & Woods, M. N. (1982). Estrogen excretion patterns and plasma levels in vegetarian and omnivorous women. *New England Journal of Medicine*, **307**, 1542–7.

Goldin, B. R., Adlercreutz, H., Gorbach, S. L., Woods, M. N., Dwyer, J. T., Conlon, T., Bohn, E. & Gershoff, S. N. (1986). The relationship between estrogen levels and diets of Caucasian American and Oriental immigrant women. *American Journal of Clinical Nutrition*, **44**, 945–53.

Goldzieher, J. W. (1989). Pharmacology of contraceptive steroids: a brief review. *American Journal of Obstetrics and Gynecology*, **160**, 1260–3.

Goldzieher, J. W. & Brody, S. A. (1990). Pharmacokinetics of ethinyl estradiol and mestranol. *American Journal of Obstetrics and Gynecology*, **163**, 2114–19.

Goldzieher, J. W., Dozier, T. S. & de la Pena, A. (1980). Plasma levels and pharmacokinetics of ethynyl estrogens in various populations. *Contraception*, **21**, 1–17.

Gray, R. H. (1980). Patterns of bleeding associated with the use of steroidal contraceptives. In *Endometrial Bleeding and Steroidal Contraception*, ed. E. Diczfalusy, I. S. Fraser and F. T. G. Webb, pp. 14–49. Bath, England: Pitman

Press, Ltd.

Guillebaud, J. (1986). *Contraception: Your Questions Answered.* New York: Churchill Livingstone.

Hill, P., Chan, P., Cohen, L., Wynder, E. & Kuno, K. (1977). Diet and endocrine-related cancer. *Cancer,* **39,** 1820–6.

Jasienska, G. & Ellison, P. T. (1993). Heavy workload impairs ovarian function in Polish peasant women. *American Journal of Physical Anthropology,* Supplement **16,** 117–18.

Johannisson, E., Landgren, B. M. & Diczfalusy, E. (1982). Endometrial morphology and peripheral steroid levels in women with and without intermenstrual bleeding during contraception with the 300 µg Norethisterone (NET) minipill. *Contraception,* **25,** 13–31.

Kappas, A., Anderson, K. E., Conney, A. H., Pantuck, E. J., Fishman, J. & Bradlow, H. L. (1983). Nutrition-endocrine interactions: Induction of reciprocal changes in the Δ4-5α-reduction of testosterone and the cytochrome P-450-dependent oxidation of estradiol by dietary macronutrients in man. *Proceedings of the National Academy of Sciences, USA,* **80,** 7646–9.

Kaufman, J. M., Thiery, M. & Vermeulen, A. (1981). Plasma levels of ethinylestradiol (EE) during cyclic treatment with combined oral contraceptives. *Contraception,* **24,** 589–602.

Key, T. J. A., Chen, J., Wang, D. Y., Pike, M. C. & Boreham, J. (1990). Sex hormones in women in rural China and in Britain. *British Journal of Cancer,* **62,** 631–6.

Koppert, G. (1988). *Alimentation et culture chez les Tamang, les Ghale et les Kami du Nepal.* Thesis Aix Marseille, France: Thèse du troisième cycle en anthropologie.

Kuhl, H. (1990). Pharmacokinetics of oestrogens and progestogens. *Maturitas,* **12,** 171–97.

Landgren, B. M. & Diczfalusy, E. (1980). Hormonal effects of the 300 µg Norethisterone (NET) minipill. 1. Daily steroid levels in 43 subjects during a pretreatment cycle and during the second month of NET administration. *Contraception,* **21,** 87–113.

Lindsay, J. (1991). The politics of population control in Namibia. In *Women and Health in Africa,* ed. M. Turshen, pp. 143–67. Trenton, NJ: African World Press.

Longcope, C., Gorbach, S., Goldin, B., Woods, M., Dwyer, J., Morrill, A. & Warram, J. (1987). The effect of a low fat diet on estrogen metabolism. *Journal of Clinical Endocrinology and Metabolism,* **64,** 1246–50.

Longcope, C. (1990). Relationship of estrogen to breast cancer, of diet to breast cancer, and of diet to estradiol metabolism. *Journal of the National Cancer Institute,* **82,** 896–7.

MacMahon, B., Cole, P., Brown, J. B., Aoki, K., Lin, T. M., Morgan, R. W. & Woo, N. (1971). Oestrogen profiles of Asian and North American women. *The Lancet,* October 23, 900–2.

MacMahon, B., Cole, P., Brown, J. B., Aoki, K., Lin, T. M., Morgan, R. W. & Woo, N. (1974). Urine oestrogen profiles of Asian and North American women. *International Journal of Cancer,* **14,** 161–7.

Musey, P. I., Collins, D. C., Bradlow, H. L., Gould, K. G. & Preedy, J. R. K. (1987).

Effect of diet on oxidation of 17β-estradiol *in vivo*. *Journal of Clinical Endocrinology and Metabolism*, **65**, 792–5.

Newburger, J. & Goldzieher, J. W. (1985). Pharmacokinetics of ethynyl estradiol: A current view. *Contraception*, **32**, 38–44.

Panter-Brick, C., Lotstein, D. S. & Ellison, P. T. (1993). Seasonality of reproductive function and weight loss in rural Nepali women. *Human Reproduction*, **8**, 684–90.

Paulsen, M. L., Varaday, A., Brown, B. W. J. & Kalman, S. M. (1974). A randomized contraceptive trial comparing a daily progestogen with a combined oral contraceptive steroid. *Contraception*, **9**, 497–507.

Prasad, K. V. S., Nair, K. M., Sivakumar, B., Prema, K. & Rao, B. S. N. (1981). Plasma levels of norethindrone in Indian women receiving norethindrone enanthate (20 mg) injectable. *Contraception*, **22**, 497–506.

Read, G. F. (1993). Status report on measurement of salivary estrogens and androgens. In *Saliva as a Diagnostic Fluid*, ed. D. Malamud and L. Tabak. New York: The New York Academy of Sciences. *Annals of the New York Academy of Sciences*, **694**, 146–60.

Rivera, R. (1993). Study and introduction of family planning methods in developing countries. *Annals of Medicine*, **25**, 57–60.

Rose, D. P., Boyar, A. P., Cohen, C. & Strong, L. E. (1987). Effect of a low-fat diet on hormonal levels in women with cystic breast disease. I. Serum steroids and gonadotropins. *Journal of the National Cancer Institute*, **78**, 623–6.

Rosetta, L. (1993). Female reproductive dysfunction and intense physical training. *Oxford Reviews of Reproductive Biology*, **15**, 113–41.

Sang, G. W., Fotherby, K., Howard, G., Elder, M. & Bye, P. G. (1981). Pharmacokinetics of norethisterone oenanthate in humans. *Contraception*, **24**, 15–27.

Seaton, B. & Riad-Fahmy, D. (1980). Use of salivary progestorone to monitor menstrual cycles in Bangladeshi women. *Journal of Endocrinology*, **87**, 21P.

Shenfield, G. M. & Griffin, J. M. (1991). Clinical pharmacokinetics of contraceptive steroids: an update. *Clinical Pharmacokinetics*, **20**, 16–37.

Shimizu, H., Ross, R. K., Berstein, L., Pike, M. C. & Henderson, B. E. (1990). Serum oestrogen levels in postmenopausal women: comparison of American whites and Japanese in Japan. *British Journal of Cancer*, **62**, 451–3.

Shultz, T. D. & Leklem, J. E. (1983). Nutrient intake and hormonal status of premenopausal vegetarian Seventh-day Adventists and premenopausal non-vegetarians. *Cancer*, **4**, 247–59.

Stadel, B. V., Sternthal, P. M., Schlesselman, J. J., Douglas, M. B., Hall, W. D., Kaul, L. & Ahluwalia, B. (1980). Variation of ethinylestradiol blood levels among healthy women using oral contraceptives. *Fertility and Sterility*, **33**, 257–60.

Stanczyk, F. Z., Lobo, R. A., Chiang, S. T. & Woutersz, T. B. (1990). Pharmacokinetic comparison of two triphasic oral contraceptive formulations containing levonorgestrel and ethinylestradiol. *Contraception*, **41**, 39–53.

Stanczyk, F. Z., Mroszczak, E. J., Ling, T., Runkel, R., Henzl, M., Miyakawa, I. & Goebelsmann, U. (1983). Plasma levels and pharmacokinetics of norethindrone and ethinylestradiol administered in solution and as tablets to women. *Contraception*, **28**, 241–51.

Talwar, P. P. & Berger, G. S. (1977). The relation of body weight to side effects associated with oral contraceptives. *British Medical Journal*, **25**, 1637–8.

Trichopoulos, D., Yen, S., Brown, J., Cole, P. & MacMahon, B. (1984). The effect of Westernization on urine estrogens, frequency of ovulation, and breast cancer. *Cancer*, **53**, 187–92.

Van der Walt, L. A., Wilmsen, E. N., Levin, J. & Jenkins, T. (1977). Endocrine studies on the San ('Bushman') of Botswana. *South African Medical Journal*, 230–2.

Van der Walt, L. A., Wilmsen, E. N. & Jenkins, T. (1978). Unusual sex hormone patterns among desert-dwelling hunter gatherers. *Journal of Clinical Endocrinology and Metabolism*, **46**, 658–63.

Vitzthum, V. J. (1990). An adaptational model of ovarian function. *Research Report of the Population Studies Center*, University of Michigan, Ann Arbor, Michigan.

Vitzthum, V. J., Ellison, P. T. & Sukalich, S. (1994). Salivary progesterone profiles of indigenous Andean women. *American Journal of Physical Anthropology Supplement*, **18**, 201–2.

Warren, M. P. (1983). Effects of undernutrition on reproductive function in the human. *Endocrine Reviews*, **4**, 363–77.

Williams, M. C. & Goldzieher, J. W. (1980). Chromatographic patterns of urinary ethynyl estrogen metabolites in various populations. *Steroids*, **36**, 255–83.

World Health Organization (1978). Multinational comparative clinical evaluation of two long-acting injectable contraceptive steroids: norethisterone oenanthate and medroxyprogesterone acetate. 2. Bleeding patterns and side effects. *Contraception*, **17**, 395–407.

World Health Organization (1982a). Multinational comparative clinical trial of long-acting injectable contraceptives: norethisterone enanthate given in two dosage regimens and depot-medroxyprogesterone acetate. A preliminary report. *Contraception*, **25**, 1–11.

World Health Organization (1982b). A randomized, double-blind study of two combined and two progestogen-only oral contraceptives. *Contraception*, **25**, 243–53.

Worthman, C. M., Jenkins, C. L., Stallings, J. F. & Lai, D. (1993). Attenuation of nursing-related ovarian suppression and high fertility in well-nourished, intensively breast-feeding Amele women of lowland Papua New Guinea. *Journal of Biosocial Science*, **25**, 425–43.

Wright, S. W., Fotherby, K. & Fairweather, F. (1970). Effect of daily small doses of norgestrel on ovarian function. *Journal of Obstetrics and Gynaecology of the British Commonwealth*, **77**, 65–8.

Yong-en, S., Chang-hai, H., Gu-Jiang & Fotherby, K. (1987). Pharmacokinetics of norethisterone in humans. *Contraception*, **35**, 465–75.

Part II
Interpopulation variability

5 Age and developmental effects on human ovarian function

P. T. ELLISON

Ever since Malthus assigned fertility regulation to the realm of social behaviour and 'moral restraint', there appears to have been a reluctance on the part of demographers to view normal (i.e. non-pathological) biological factors as presenting anything more than some upper limit on human fecundity. Menarche, menopause, the fixed duration of gestation, perhaps some unavoidable frequency of pregnancy loss, these are often viewed as providing a set of bounds on the human potential for offspring production (Bongaarts & Potter, 1983; Campbell & Wood, 1988). But an understanding of variation within these bounds is ordinarily sought in terms of human behaviour; rates of marriage and divorce, the use of contraception and abortion, the frequency of intercourse and spousal separations, lactation and weaning. There is no question that these aspects of our social and personal behaviour strongly influence our fertility, nor that they themselves are shaped by more distal elements of our cultural values and social and economic institutions. At the same time it is becoming increasingly apparent that biologically and ecologically mediated variation in female fecundity also contributes to patterns of human fertility. Age effects on female fecundity provide a particularly clear example both of the importance of biological factors affecting fertility, and of the potential for biological anthropologists to contribute empirically and theoretically to our understanding of human fertility variation.

Age and natural fertility

The age pattern of female fertility is one of the most familiar aspects of human demography and one of the basic elements in our understanding of human population dynamics. Yet it is only relatively recently that the possibility of an underlying age pattern of female fecundity has been appreciated (Wood, 1989; Ellison, 1990; Weinstein et al., 1990). This is perhaps the more remarkable given the strong evidence of similarity in the

69

age patterns of female marital fertility among natural fertility populations that has been noted ever since the seminal work of Louis Henry (1961). Henry defined natural fertility as the absence of conscious parity-specific fertility regulation. Although controversial (Blake, 1985; Wilson, Oeppen & Pardoe, 1988), this definition has been extremely heuristic, particularly in focusing attention on the rate at which births occur during the reproductive span rather than on factors influencing the starting and stopping of childbearing. It was also an operational definition that allowed Henry to empirically distinguish natural from controlled fertility populations by their parity progression ratios. When he did so, he made two important observations concerning natural fertility variation: (1) that the *level* of age-specific marital fertility varies quite appreciably between different populations, and (2) that the *pattern* of age-specific marital fertility is none the less remarkably constant. Henry's analysis led him to suggest that the explanation for the first observation might lie in the variable effect of lactation in delaying the resumption of ovulation postpartum. He had less to say about the second observation, except that the declining fertility of women over 30 years of age was a consequence both of increasing numbers of infertile women with age and of increasing birth intervals among those who remain fertile.

Evidence accumulated in succeeding years has only confirmed both of Henry's observations (Howell, 1979; Bongaarts & Potter, 1983; Brecken-ridge, 1983; Knodel, 1983; Menken, Trussell & Larsen, 1986; Campbell & Wood, 1988; Wilson *et al.*, 1988; Wood, 1989, 1990). The similarity in age patterns of female natural fertility between populations is apparent when age-specific marital fertility rates are standardized on the rate for the 20–24 year-old age class in each population (Howell, 1979; Knodel, 1983; Wood, 1989). The resulting pattern is roughly parabolic (Fig. 5.1). Fertility rates rise steadily over the first decade of the reproductive span (Bendel & Hua, 1978), reaching a peak in the third decade of life. Fertility levels in the 20–24 and 25–29 year age classes are often nearly equal, with the highest levels occurring in the latter age class in many populations (Knodel, 1983; Wood, 1989). During the fourth decade of life fertility levels begin a steady decline lasting a decade or more preceding menopause. It should be noted, however, that this trajectory characterizes cross-sectional data. The longitudinal trajectories implied by the reproductive histories of individual women may be much more rectangular, beginning and ending more abruptly.

Explanations for the age pattern of natural fertility often invoke the role of behavioural changes, particularly declining frequency of intercourse with age (James, 1979, 1981, 1983; Ruzicka & Bhatia, 1982; Udry, Devon & Coleman, 1982), and the onset of permanent sterility (Henry, 1961;

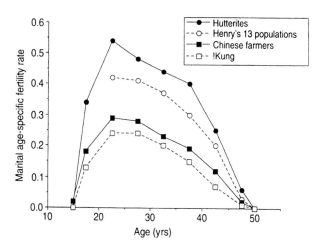

Fig. 5.1. Age patterns of female marital fertility in representative natural fertility populations. (Redrawn from Howell, 1979.)

Menken *et al.*, 1986; Willson *et al.*, 1988). Recent quantitative models indicate, however, that changing rates of sexual intercourse make only minor contributions to age-specific declines in natural fertility after age 30, and cannot account at all for the increasing infertility prior to age 25 (Wood & Weinstein, 1988; Weinstein *et al.*, 1990). Neither do increasing rates of permanent sterility with age account for increasingly long waiting times to conception among those still fertile (Henry, 1961; Menken *et al.*, 1986). Rather, new evidence from both clinical and field studies suggests that age-specific changes in female reproductive physiology contribute significantly to the characteristic age pattern of natural fertility.

Clinical evidence of age patterns in female fecundity

The most compelling clinical evidence for age-specific changes in female fecundity has come from infertility treatment programs, where cumulative data consistently depict a pattern of declining success rates with age in women past the mid-thirties (Table 5.1). Different studies control for different potentially confounding effects, but most control for frequency of intercourse and male fecundity, either by relying on timed artificial insemination by donor (AID), or on *in vitro* fertilization (IVF). Fédération CECOS (1982) report on success rates in AID for 2193 women in France, noting that success rates in establishing a pregnancy begin to decline

Table 5.1. *Clinical studies demonstrating a decline in female fecundity with advancing age*

Artificial insemination by donor (AID)	*Ovulation induction*
Fédération CECOS, 1982	Levran *et al.*, 1991
Virro & Shewchuk, 1984	Navot *et al.*, 1991
Edvinsson *et al.*, 1990	Abdalla *et al.*, 1993
	Meldrum, 1993
	Flamigni *et al.*, 1993
In vitro fertilization (IVF)	*Ovum donation*
Sharma *et al.*, 1988	Pearlstone *et al.*, 1992
Corson *et al.*, 1989	McClure *et al.*, 1993
Hughes, King & Wood, 1989	Dickey *et al.*, 1993*a, b*
Toner *et al.*, 1991	
FIVNAT, 1993	

significantly in women over age 30 and increasingly in women over age 35. Subsequent studies of women undergoing AID elsewhere have confirmed this finding (Virro & Shewchuk, 1984; Edvinsson *et al.*, 1990). Similarly, success rates in establishing ongoing pregnancies have been found to vary inversely with age in women undergoing IVF (Padilla & Garcia, 1989; Toner *et al.*, 1991; FIVNAT, 1993; Abdalla *et al.*, 1993), ovulation induction (Pearlstone *et al.*, 1992, McClure *et al.*, 1993; Dickey *et al.*, 1993*a,b*), and ovum donation (Sauer, Paulson & Lobo, 1990, Levran *et al.*, 1991; Navot *et al.*, 1991; Meldrum, 1993; Flamigni *et al.*, 1993). The single largest study is the report of cumulative results from more than 77 000 IVF procedures carried out in France between 1986 and 1990 (FIVNAT, 1993). Like the Fédération CECOS data, the FIVNAT data show a progressive decline in success rates with female age, detectable as early as age 30 and accelerating after age 35.

Several attempts have been made to use the data from infertility treatment programs to localize the anatomical source of declining female fecundity with age. Ovum donation provides particularly useful data in this regard (Levran *et al.*, 1990; Navot *et al.*, 1991; Rotsztejn & Asch, 1991), since the age of the oöcytes and of the uterus can vary independently. The data must be interpreted with caution, however, since the hormonal milieu of both the donor and the recipient is manipulated exogenously. Some, but not all, aspects of age-related variation in female fecundity may thus be revealed by analysing success rates in this procedure.

The fact that successful pregnancies can be established through ovum donation even in post-menopausal women indicates that uterine age is not ultimately limiting on female fecundity: given appropriate hormonal stimulation and a viable embryo the ovary of a post-menopausal woman is capable of successful implantation and gestation (Sauer *et al.*, 1990; Sauer, Paulson & Lobo, 1992, 1993). Success rates in establishing pregnancies via ovum donation decline with recipient age, however, indicating some reduction in the capability of the endometrium to implant and sustain an embryo (Levran *et al.*, 1991; Meldrum, 1993; Flamigni, 1993). Meldrum (1993) reports that the lower success rate in recipients over 40 years compared to those under 40 years can be corrected by doubling the dose of exogenous progesterone administered, suggesting possible age-related changes in ovarian endocrine function or endometrial receptor density may be involved.

Navot *et al.* (1991) and Abdalla *et al.* (1993) report that pregnancy success rates for ovum donation in women over 40 years are higher than success rates for IVF in similarly aged women using their own ova, and interpret the data as indicating a significant effect of age on oöcyte quality. Flamigni *et al.* (1993) compare the pregnancy success rates in women undergoing ovum donation to the success rates of their specific donors undergoing IVF with eggs from the same harvested cohort. They find a significant decline in success rates with the age of recipients and interpret the results as demonstrating an effect of uterine age. In a design in which both the age of the recipients (over and under 33 years) and the age of the donors (over and under 32 years) in an ovum donation program were analysed, Levran *et al.* (1991) find that recipient age affects the probability of a pregnancy being initially established, while donor age affects the probability of successful continuance of those pregnancies which are established. Taken together, the results from ovum donation programmes seems to indicate that both the age of the recipient (ovarian and uterine age) and the age of the donor (ovarian and oöcyte age) have a negative effect on the probability of a successful pregnancy. It should be noted that, although donor age is often interpreted as equivalent to oöcyte age, it cannot be distinguished in these designs from the age of the ovary and its effect on the development and maturation of the oöcyte. In particular, it should be remembered that the granulosa cells making up each primordial follicle, responsible for oestradiol production and nurture of the ovum, are themselves as old as the oöcyte they contain. Nor can 'uterine age' effects in recipients be separated fully from the potential effects of lower ovarian steroid stimulation over preceding cycles.

Basal follicle-stimulating hormone (FSH) levels have been used as an index of 'ovarian age' inversely reflecting the remaining follicular supply of the ovary (Toner *et al.*, 1991; Pearlstone *et al.*, 1992). A smaller follicular reserve is assumed to result in less negative feedback (by oestradiol and/or inhibin) on FSH release by the pituitary. Both chronological age and basal FSH have been found to contribute independently to success rates in IVF (Toner *et al.*, 1991) and ovulation induction (Pearlstone *et al.*, 1992). McClure *et al.* (1993) report that both chronological age and basal, unstimulated oestradiol levels are predictive of success in ovulation induction. The predictive value of follicular oestradiol may be related to the independently demonstrated importance of endometrial thickness in predicting success in ovulation induction (Shoham *et al.*, 1991, Dickey *et al.*, 1993*a,b*). From these data it appears that the functional capacity of the ovary in its endocrine role, in addition to the quality of the oöcytes themselves, may be compromised by increasing age.

Declining female fecundity at older ages has received considerable attention because it is linked to a tremendous demand for clinical services. Less clinical attention has been given to the issue of increasing fecundity over the first decade of life. The notion of 'adolescent sterility' is often misused in this context to imply the existence of some relatively brief (i.e. a few years) lag time between the onset of menstruation and a similarly discrete 'onset' of fecundity. As will be noted below, physiological studies indicate a much more gradual and sustained trajectory of increasing female fecundity with age, matching in its duration the rising phase of natural fertility patterns.

Evidence of age patterns in ovarian function

Inferences about age variation in ovarian function, or the lack of it, are often incorrectly made in the demographic literature from data on menstrual function (Bongaarts & Potter, 1983; Wood, 1989). These inferences are erroneous in that they fail to recognize that variation in menstrual function represents only a narrow and extreme portion of the full range of variation in ovarian function (Fig. 5.2) (Ellison, 1990, 1991). Because menstruation is a qualitative event, one that can be scored as present or absent, and because it is easy to observe and report, more attention has been devoted to menstruation as an indicator of fecundity that it may deserve. While it is almost certainly true that an amenorrheic woman is infecund, and that an oligomenorrheic woman has low fecundity, it is not true that regular menstruation indicates full or homogeneous fecundity (Lenton, Gelsthorp & Harper, 1988; Weinstein, Wood & Greenfield, 1993).

The impression of a lack of significant age variation in ovarian function has derived its most important support from the work of Treloar and

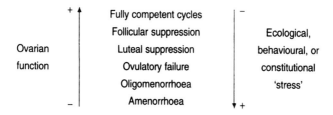

Fig. 5.2. The continuum of ovarian function (Ellison, 1990, 1991).

colleagues. Treloar *et al.* (1967) present over 25 000 person years of data on menstrual cycle variability gathered from over 2000 women indicating longer and more irregular cycle lengths in the years immediately after menarche and immediately before menopause, but with very little variability over the majority of the reproductive span from 20 to 40 years of age. Equally impressive, however, but less widely cited, is work by Döring indicating a more parabolic and less rectangular trajectory of ovarian function with age. Döring (1969) analyses basal body temperature recordings, an indirect reflection of progesterone secretion and the presence of a corpus luteum, representing 3264 person–months of observation on 481 women, concluding that ovulatory frequency and rates of luteal sufficiency increase with age after menarche until the mid-twenties and begin to decline as early as the mid-thirties. Direct endocrinological investigations of age variation in ovarian function have been few and limited in scope. Several studies have confirmed the high frequency of anovulatory and luteally insufficient cycles in the perimenarcheal and perimenopausal periods (Sherman & Korenman, 1975; Sherman, West & Korenman, 1976; Apter & Vihko, 1978; Metcalf, Donald & Livesey, 1981; Metcalf *et al.*, 1983, Lenton *et al.*, 1984; Vihko & Apter, 1984) but have not addressed the issue of variation in ovarian function between these extremes.

This issue has been addressed in a study of salivary progesterone levels among 124 healthy, regularly menstruating women from the Boston area between the ages of 18 and 44, all of whom were of stable weight, within normal ranges of weight for height, and none of whom exercised regularly or used oral contraception (Lipson & Ellison, 1992, 1994). The criteria of subject selection were construed to specifically exclude women in the perimenarcheal or perimenopausal periods, or women with anything other than consistent intermenstrual intervals falling within the normal range of cycle lengths established by Treloar *et al.* (1967). Saliva samples were collected daily by all subjects and analysed for progesterone content by

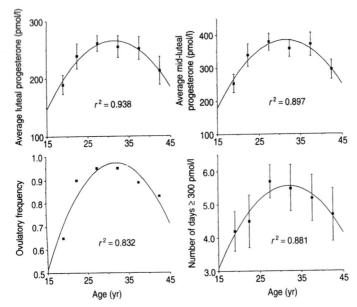

Fig. 5.3. Mean ($\pm$SE) values of four indices of luteal function by age for
124 Boston women between the ages of 18 and 45 derives from daily salivary
progesterone levels (Lipson & Ellison, 1992). The indices are: average luteal
progesterone over the 16 days prior to menstrual onset; average mid-luteal
progesterone from 5 to 9 days prior to menstrual onset; ovulatory frequency,
defined by the presence of at least one luteal value ≥ 300 pmol/l; and the total
number of luteal days with values ≥ 300 pmol/l. Solid lines are the best fit
second degree polynomial regressions, with r-squared values given in the
Figure.

established procedures (Ellison, Peacock & Lager, 1986; Ellison, 1988,
1993). Analysis of the resulting profiles indicates significant age variation in
luteal activity, levels increasing with age between 18 and 24 years, and
declining with age after 35 years. Average values for several indices of luteal
function (ovulatory frequency, average luteal progesterone, average mid-
luteal progesterone, and the number of luteal days with values ≥ 300
pmol/l) all show parabolic trajectories across the age groups studied, and
are well fit by second-order polynomial regressions (linear and quadratic
terms and overall regressions significant at the 0.05 level in all cases, r^2
values between 0.86 and 0.95, Fig. 5.3). This is to date the only endocrine
study of age variation in ovarian function between the extremes of the
perimenarcheal and perimenopausal periods. On the basis of these results
we have recently generated the first age-stratified reference values for the
clinical assessment of luteal function (Lipson & Ellison, 1994).

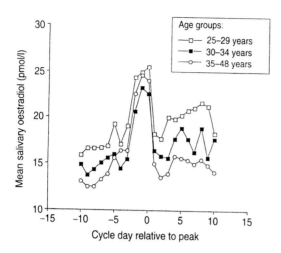

Fig. 5.4. Average profiles of salivary oestradiol for three age groups of Boston women (O'Rourke & Ellison, 1993*a*).

The crucial importance of luteal progesterone to the success of implantation and the maintenance of ongoing pregnancy has been underscored by the effectiveness of the synthetic progesterone antagonist, RU486, in terminating pregnancy and inducing menstruation at any time from the early luteal phase through the late second trimester (Baulieu, 1989). The dose responsiveness of female fecundity to luteal progesterone levels has also been documented in the context of assisted reproduction procedures (Meldrum, 1993). Short luteal phases, low luteal progesterone production, and luteal phases which are poorly synchronized with follicular development and gamete maturation are all associated with low fecundity and high rates of embryonic loss (Liu *et al.*, 1988; Maslar, 1988; Stouffer, 1988; McNeely & Soules, 1988). Comparison of conception and non-conception cycles indicates that conception cycles which result in ongoing pregnancies are associated with higher progesterone both before and after implantation, and longer duration luteinizing hormone surges, than are non-conception cycles or those that result in embryonic loss (Lenton *et al.*, 1988).

A similar study of age variation in salivary oestradiol profiles has been conducted in 53 healthy, regularly menstruating, Boston area women between the ages of 24 and 48, meeting the same subject criteria as in the Lipson and Ellison study (45 of the women were subjects in both studies) (O'Rourke & Ellison, 1993*a,b*). Average follicular and average luteal oestradiol values also decline with increasing subject age, with significant differences apparent in this case as early as age 30 (Fig. 5.4). Low follicular oestradiol is correlated with absolutely smaller follicular size (Apter *et al.*, 1987), low oöcyte fertilizability (Yoshimura & Wallach, 1987), reduced endometrial thickness (Shoham *et al.*, 1991), and low pregnancy rates in ovulation induction (McClure *et al.*, 1993). Luteal phase oestradiol is important in inducing progesterone receptors in the endometrium (Fritz, Westfahl & Graham, 1987; Baulieu, 1990) and in regulating pituitary gonadotropin secretion (de Ziegler *et al.*, 1992). Declining oestradiol profiles with increasing age thus contribute to lower female fecundity in several ways.

   The existence of significant age variation in ovarian function makes it important to control for age in studies of other variables that might also affect female reproductive physiology, such as lactation, nutrition or disease (Nath *et al.*, 1994). Controlling for age is also important when making comparisons of ovarian function or fecundity between populations (Ellison, Peacock & Lager, 1989; Ellison, 1990, 1993, 1994; Panter-Brick, Lotstein & Ellison, 1993; Ellison *et al.*, 1993*a*). In order to appropriately exercise such control, however, the degree to which age variation in female reproductive physiology is comparable between different populations must be considered. At least two studies, one in New Guinea (Wood, Johnson & Campbell, 1985), and one in Africa (Phillip, Worthman & Stallings, 1991), have seemed to indicate that female fecundity in those populations may not decline with age, or may even increase.

Population comparisons of age variation in ovarian function

Directly comparable data have been collected on variation in salivary progesterone profiles that allow some analysis of age variation from four populations: middle-class Boston women, Lese horticulturalists from the Ituri Forest of Zaïre, Tamang agro-pastoralists from the Himalayan foothills of Nepal, and Quechua-speaking women from highland Bolivia, widely separated geographically, genetically, ecologically, and culturally. Data from Boston women has already been described above. Data from Zaïre cover 144 cycles from Lese women collected in 1984 and 1989 (Ellison *et al.*, 1989, Bailey *et al.*, 1992). Some women are represented in the sample from both field seasons at ages 5 years apart. Data from Nepal include one cycle each from 45 women collected in 1990 (Panter-Brick *et al.*, 1993). Data from Bolivia are from a preliminary study of 18 women, each of which is represented by a single cycle (Vitzthum *et al.*, in prep.). Subjects in Zaïre and Nepal met the same criteria of cycle regularity as the Boston

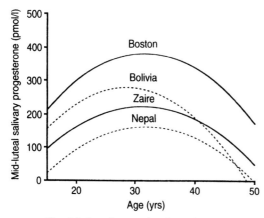

Fig. 5.5. Best fit second order polynomial regressions of mid-luteal progesterone levels on age for four populations (Ellison *et al.*, 1993*b*; Vitzthum *et al.*, in prep).

women. The Bolivian sample was not so constrained and includes some older women with irregular menstruation. For this reason, and because of the small sample, the data from Bolivia are presented for illustrative purposes only and are not included in some analyses.

A two-way analysis of variance of the mid-luteal progesterone levels from Boston, Zaïre, and Nepal by population and age group across seven five-year age groups from 15 to 50 indicates highly significant differences between age groups in all three populations ($p < 0.001$), and between populations at each age ($p < 0.001$), but no significant interaction effect ($F = 0.35$, $p = 0.98$). The near perfect parallelism of the age trends can be visualized by comparing the best fitting second-order polynomial regressions obtained for each population (Fig. 5.5). The best fitting curve to the Bolivian data is also represented in Fig. 5.5, exhibiting a steeper slope at older ages than the other three populations. It is likely that this is a result of the more inclusive criteria for subject selection in this pilot study allowing the inclusion of women already in the menopausal transition.

Taken as a whole, the data from these populations indicate a remarkable degree of similarity in the age patterns of ovarian function as reflected by salivary progesterone levels. The pattern described in detail for Boston women appears to hold for women from Zaïre and Nepal as well. The Bolivian data are useful in indicating the potential for different patterns to emerge in this analysis. Whether the Bolivian pattern actually departs from the pattern of the other populations, however, cannot be answered without more data from subjects meeting the similar criteria of menstrual regularity.

Developmental effects on adult ovarian function

The consistency of the age pattern of luteal function across such genetically, ecologically, and culturally distinct populations strongly suggests that it represents a general feature of human female reproductive physiology largely unaffected by environmental circumstances. The overall level of luteal function, however, varies markedly between populations (Ellison et al., 1993b). In fact, as reflected in salivary progesterone levels, ovarian function displays the same two features, variance in level between populations and consistency in age pattern across populations, that Henry originally observed in natural fertility. As is also true of natural fertility, the age pattern and level of ovarian function appear to be determined independently. Certainly, various acute factors, such as energy balance, are capable of modulating adult ovarian function, as we have demonstrated in Boston (Ellison & Lager, 1985, 1986; Lager & Ellison, 1990), Zaïre (Ellison et al., 1989; Bailey et al., 1992), and Nepal (Panter-Brick et al., 1993). The range of variation ascribable to these factors, however, does not appear to encompass the variation in average levels of ovarian function between populations. Rather, the levels of ovarian function in Fig. 5.5 appear to reflect a baseline or central tendency for each population about which both age and acute variations occur.

The explicit hypothesis that baseline levels of adult ovarian function might be established during the period of growth and development was introduced in two earlier publications (Ellison, 1990, hypothesis 8; Ellison, 1991, hypothesis 4). This hypothesis is motivated both by analogy with other aspects of human anatomy and physiology which show similar developmental effects, and by the theoretical expectation that chronic stresses will elicit responses similar in direction but longer in duration and less reversible than acute stresses of the same sort. In this case, the hypothesis holds that some of the same conditions associated with acute suppression of ovarian function, e.g. low energy availability, are associated with slow growth and late maturation when they occur chronically, and may under those circumstances also be associated with low baseline levels of adult ovarian function.

This hypothesis properly requires longitudinal data in order to be tested. Given that adequate cross-sectional data on human ovarian function have only recently become available, it is perhaps not surprising that few

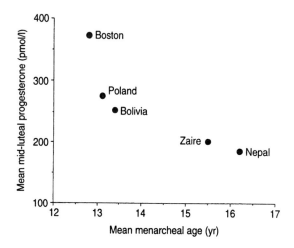

Fig. 5.6. Relationship of mean mid-luteal progesterone level for women aged 25 to 34 years to mean menarcheal age (Eveleth & Tanner, 1990) for five populations.

longitudinal data as yet exist. Those longitudinal data that do exist, however, largely support the hypothesis. Apter and Vihko (1983) have documented that early and late maturing Swedish girls have very different trajectories of ovarian function, with early maturers (menarche before age 12) attaining an ovulatory frequency of nearly 80% within 2 years of menarche and 100% within 5 years, while late maturers (menarche after age 14) only attain an ovulatory frequency of 25% within 2 years of menarche and only 50% within 5 years. Venturoli *et al.* (1987) find that late maturers continue to have a higher frequency of endocrine and menstrual abnormalities than early maturers well into adulthood, while Gardner and Valadian (Gardner, 1983; Gardner & Valadian, 1983) report a higher frequency of oligomenorrhea and dysmenorrhea among late maturers than early maturers as adults.

If the developmental hypothesis were true, one might expect a negative relationship between average menarcheal age and average level of adult ovarian function across populations. The sample size for such a comparison is at present very small. To the four populations mentioned above can be added data from rural southern Poland, collected as part of a study of energetics and ovarian function (Jasienska & Ellison, 1993). Comparing mean mid-luteal progesterone levels from these populations with reported values for mean menarcheal age collected by others (Eveleth & Tanner, 1990) reveals the expected negative correlation (Fig. 5.6). Scant confidence can be put in this relationship, to be sure, until comparable data on a larger number of populations is available. It does seem, however, that there are sufficient grounds for entertaining the developmental hypothesis and gathering additional data.

There are also good alternative hypotheses. Peacock (1990, 1991) and Vitzthum (1990) have argued that chronic energy shortage should have no effect on female fecundity, that suppression of fecundity is only adaptive as an acute response to a temporarily poor probability of a successful reproductive outcome. If the probability of success is unlikely to improve, they argue, there is no benefit to suppressing fecundity. The hypothesis I have put forward is based, however, not only on the probability of a successful reproductive outcome, but on the cost of reproduction to the mother. Evidence from the Gambia suggests that women may lower their own basal metabolisms to meet the energetic demands of reproduction (Prentice & Whitehead, 1987) and that one consequence may be increased susceptibility to respiratory and gastrointestinal diseases (Prentice *et al.*, 1983). Detailed studies also indicate that behavioural and physiological energy sparing mechanisms are inadequate in themselves to balance the energy requirements of reproduction (Prentice & Whitehead, 1987; Prentice & Prentice, 1990). Data from New Guinea (Tracer, 1991) suggest that inadequate spacing of births may result in progressive maternal depletion with consequent higher morbidity and mortality risk. Lowering fecundity by lowering ovarian function increases the waiting time to conception, extending the time a woman in a natural fertility population spends 'metabolizing for one' and decreasing the relative proportion of time spent in gestation and lactation, 'metabolizing for two'. Because the energy investment in reproduction is relatively inelastic and occurs in discrete quanta, and because other aspects of the maternal energy budget do not provide adequate potential for energy savings, the only effective strategy for maintaining long-term energy balance hinges on spacing births more widely and increasing the proportion of time spent in less energetically demanding reproductive states.

Conclusions

The fact that the age pattern of ovarian function is so consistent across human populations, despite dramatic variation in geography, ecology, and behaviour, suggests that it is a general feature of human life history. If it is correct to interpret the level of ovarian function as a physiological index of reproductive effort, the ratio of time and effort devoted to reproduction versus time and effort devoted to maintenance, then the age pattern of ovarian function could be summarized as a decade or so of increasing reproductive effort, a decade or so of sustained reproductive effort, and a decade or so of declining reproductive effort. For such a pattern to be invariant implies that it has broad functional significance, independent of specific context. How are we to understand that significance?

Early in the reproductive career the important evolutionary trade-off is likely that between current and future reproduction. Resources invested in maintenance can increase the probability of future reproduction by increasing the chances of surviving to reproduce. As a woman ages and her reproductive value declines, this benefit of investing in maintenance declines. It is, in fact, a common prediction of life history theory that reproductive effort should generally increase with age (Roff, 1992). In this way we can perhaps account for the gradual increase in female fecundity at young ages to some optimal level for a given environment, and its maintenance at that high level through the middle years of the reproductive career.

Late in reproductive life the important evolutionary trade-off is likely to be a different one, not between current and future reproduction, but between current and past reproduction. That is, the basic allocation decision may be between investing in another newborn, or continuing investment in children already born. Especially given the declining quality of zygotes as maternal age advances past the mid-thirties, a consequence not of patterns of investment or past reproduction but rather a reflection of diminishing öocyte supply, the balance may tip increasingly away from new reproductive opportunities towards continuing parental investment. This continuing investment involves the continued survival of the mother, hence a shift in favour of maintenance over reproduction, not in order to ensure future reproduction, but to ensure continued ability to invest in children already born. In this sense, declining fecundity at older ages is not a mirror image of increasing fecundity at younger ages.

Malthus is certainly right that we cannot rely on natural forces to regulate human populations. But he is almost certainly wrong in implying that biological variation in human fecundity is non-existent or unimportant. Understanding that the secular trend in growth brings with it not only earlier reproductive maturation but more rapid attainment of higher fecundity may be important in our efforts to understand teenage sexuality and pregnancy. Understanding the pattern of fecundity decline at older ages may be important in order for women to make informed reproductive and career decisions. The relationship between age at diagnosis and prognosis of breast cancer may be related to age variation in ovarian steroid production (O'Rourke & Ellison, 1993b). Population variation in the incidence of breast cancer itself may be related to population variation in baseline levels of ovarian function. If those baselines are determined during childhood and adolescence, then early interventions may be possible that will lower lifetime breast cancer risk.

The practical benefits of improving our understanding of our own reproductive biology are legion. So are the opportunities for biological anthropologists to contribute to that understanding. Our disciplinary focus on normal variation, our global perspective on the range of that variation, our orientation toward field rather than clinical research, and our acceptance of the guiding framework of evolutionary theory set us apart from demographers, epidemiologists, and clinical scientists. Bringing that perspective to bear on the study of human fertility may not only add to our understanding of our own biology and evolutionary past, but also may contribute significantly to our well-being and hopes for the future.

ACKNOWLEDGEMENTS

I thank all the members of the Reproductive Ecology Laboratory at Harvard University, past and present, especially Richard Bribiescas, Ben Campbell, Drew Colfax, Marion Eakin, Diana Graham, Grazyna Jasienska, Cheryl Knott, Catherine Lager, Susan Lipson, Deborah Lotstein, Mary O'Rourke, Diana Sherry, Sara Sukalich, and my collaborators in the field, especially Robert Bailey, Gillian Bentley, Alisa Harrigan, Mark Jenike, Catherine Panter-Brick, Nadine Peacock, and Virginia Vitzthum for their many substantive contributions to the work reviewed here. I also thank the women of Boston, Zaïre, Nepal, Poland, and Bolivia who have graciously aided us in our studies. This work has been supported by the National Science Foundation, Washington, DC.

REFERENCES

Abdalla, H. I., Burton, G., Kirkland, A., Johnson, M. R., Leonard, T., Brooks, A. A. & Studd, J. W. W. (1993). Age, pregnancy, and miscarriage: uterine versus ovarian factors. *Human Reproduction*, 8, 1512–17.

Apter, D., Raisanen, I., Ylostalo, R. & Vihko, R. (1987). Follicular growth in relation to serum hormonal patterns in adolescents compared with adult menstrual cycles. *Fertility and Sterility*, **47**, 82–8.

Apter, D. & Vihko, R. (1978). Hormonal pattern of adolescent menstrual cycles. *Journal of Clinical Endocrinology and Metabolism*, **47**, 944–54.

Apter, D. & Vihko, R. (1983). Early menarche, a risk factor for breast cancer, indicates early onset of ovulatory cycles. *Journal of Endocrinology and Metabolism*, **57**, 82–6.

Bailey, R. C., Jenike, M. R., Bentley, G. R., Harrigan, A. M. & Ellison, P. T. (1992). The ecology of birth seasonality among agriculturalists in central Africa. *Journal of Biosocial Science*, **24**, 393–412.

Baulieu, E.-E. (1989). Contragestion and other clinical applications of RU486, an antiprogesterone at the receptor. *Science*, **245**, 1351–7.

Baulieu, E.-E. (1900). Hormones, a complex communication network. In *Hormones*, ed. E.-E. Baulieu and P. A. Kelly, pp. 3–169. New York: Routledge, Chapman and Hall.

Bendel, J. P. & Hua, C. (1978). An estimate of the natural fecundability ratio curve. *Social Biology*, **25**, 210–27.

Blake, J. (1985). The fertility transition: continuity or discontinuity with the past? International Population Conference, Florence 1985. Liege: IUSSP, Vol 4, pp. 393–405.

Bongaarts, J. & Potter, R. G. (1983). *Fertility, Biology, and Behavior: An Analysis of the Proximate Determinants*. New York: Academic Press.

Breckenridge, M. B. (1983). *Age, Time, and Fertility*. New York: Academic Press.

Campbell, K. L. & Wood, J. W. (1988). Fertility in traditional societies. In *Natural Human Fertility and Biological Determinants*, ed. P. Diggory, M. Potts, S. Teper, pp. 39–69, London: Macmillan.

Corson, S. L., Dickey, R. P., Gocial, B., Batzer, F. R., Eisenberg, E., Huppert, L. & Maislin, G. (1989). Outcome in 242 *in vitro* fertilization – embryo replacement or gamete intrafallopian transfer-induced pregnancies. *Fertility and Sterility*, **51**, 644–50.

de Ziegler, D., Bergeron, C., Cornel, C., Medalie, D. A., Massai, M. R., Milgrom, E., Frydman, R., Bouchard, P. (1992). Effects of luteal estradiol on the secretory transformation of human endometrium and plasma gonadotropins. *Journal of Clinical Endocrinology and Metabolism*, **74**, 322–31.

Dickey, R. P., Olar, T. T., Taylor, S. N., Curole, D. N. & Harrigill, K. (1993*a*). Relationship of biochemical pregnancy to pre-ovulatory endometrial thickness and pattern in patients undergoing ovulation induction. *Human Reproduction*, **8**, 327–30.

Dickey, R. P., Olar, T. T., Taylor, S. N., Curole, D. N. & Matulich, E. M. (1993*b*). Relationship of endometrial thickness and pattern to fecundity in ovulation induction cycles: effect of clomiphene citrate alone and with human menopausal gonadotropin. *Fertility and Sterility*, **59**, 756–60.

Döring, G. K. (1969). The incidence of anovular cycles in women. *Journal of Reproduction and Fertility, Supplement*, **6**, 77–81.

Edvinsson, A., Forssman, L., Milson, I. & Nordfors, G. (1990). Factors in the infertile couple influencing the success of artificial insemination with donor semen. *Fertility and Sterility*, **53**, 81–7.

Ellison, P. T. (1988). Human salivary steroids: methodological considerations and applications in physical anthropology. *Yearbook of Physical Anthropology*, **31**, 115–42.

Ellison, P. T. (1990). Human ovarian function and reproductive ecology: new hypotheses. *American Anthropologist*, **92**, 933–52.

Ellison, P. T. (1991). Reproductive ecology and human fertility. In *Biological*

Anthropology and Human Affairs, ed. Lasker, G. W. and Mascie-Taylor, N., pp. 14–54. Cambridge: Cambridge University Press.

Ellison, P. T. (1993). Measurements of salivary progesterone, *Annals of the New York Academy of Sciences*, **694**, 161–76.

Ellison, P. T. (1994). Salivary steroids and natural variation in human ovarian function. *Annals of the New York Academy of Sciences*, in press.

Ellison, P. T. & Lager, C. (1985). Exercise-induced menstrual disorders. *New England Journal of Medicine*, **313**, 825–6.

Ellison, P. T. & Lager, C. (1986). Moderate recreational running is associated with lowered salivary progesterone profiles in women. *American Journal of Obstetrics and Gynecology*, **154**, 1000–3.

Ellison, P. T., Lipson, S. F., O'Rourke, M. T., Bentley, G. R., Harrigan, A. M., Panter-Brick, C. & Vitzthum, V. J. (1993a). Population variation in ovarian function. *Lancet*, **342**, 433–4.

Ellison, P. T., Panter-Brick, C., Lipson, S. F. & O'Rourke, M. T. (1993b). The ecological context of human reproduction. *Human Reproduction*, **8**, 2248–58.

Ellison, P. T., Peacock, N. R. & Lager, C. (1986). Salivary progesterone and luteal function in two low-fertility populations of northeast Zaire. *Human Biology*, **58**, 473–83.

Ellison, P. T., Peacock, N. R. & Lager, C. (1989). Ecology and ovarian function among Lese women of the Ituri Forest, Zaire. *American Journal of Physical Anthropology*, **78**, 519–26.

Eveleth, P. B. & Tanner, J. M. (1990). *Worldwide Variation in Human Growth. 2nd ed.*, Cambridge: Cambridge University Press.

Federation CECOS, F., Schwartz, D. & Mayaux, M. J. (1982). Female fecundity as a function of age. *New England Journal of Medicine*, **306**, 404–6.

FIVNAT (1993). French national IVF registry: analysis of 1986 to 1990 data. *Fertility and Sterility*, **59**, 587–95.

Flamigni, C., Borini, A., Violini, F., Bianchi, L. & Serrao, L. (1993). Oocyte donation: comparison between recipients from different age groups. *Human Reproduction*, **8**, 2088–92.

Fritz, M. A., Westfahl, P. K. & Graham, R. L. (1987). The effect of luteal phase estrogen antagonism on endometrial development and luteal function in women. *Journal of Endocrinology and Metabolism*, **65**, 1006–13.

Gardner, J. (1983). Adolescent menstrual characteristics as predictors of gynaecological health. *Annals of Human Biology*, **10**, 31–40.

Gardner, J. & Valadian, I. (1983). Changes over thirty years in an index of gynaecological health. *Annals of Human Biology*, **10**, 41–55.

Henry, L. (1961). Some data on natural fertility. *Eugenics Quarterly*, **8**, 81–91.

Howell, N. (1979). *Demography of the Dobe !Kung*, New York: Academic Press.

Hughes, E. G., King, C. & Wood, E. C. (1989). A prospective study of prognostic factors in *in vitro* fertilization and embryo transfer. *Fertility and Sterility*, **51**, 838–00.

James, W. H. (1979). The causes of the decline in fecundability with age. *Social Biology*, **26**, 330–4.

James, W. H. (1981). Distributions of coital rates and of fecundability. *Social Biology*, **28**, 334–41.

James, W. H. (1983). Decline in coital rates with spouses' ages and duration of marriage. *Journal of Biosocial Science*, **15**, 83–7.

Jasienska, G. & Ellison, P. T. (1993). Heavy workload impairs ovarian function in Polish peasant women. *American Journal of Physical Anthropology Supplement*, **16**, 117–18.

Knodel, J. (1983). Natural fertility: age patterns, levels, and trends. In *Determinants of Fertility in Developing Countries, Vol I*, ed. R. A. Bulatao and R. D. Lees, New York: Academic Press.

Lager, C. & Ellison, P. T. (1990). Effect of moderate weight loss on ovarian function assessed by salivary progesterone measurements. *American Journal of Human Biology*, **2**, 303–12.

Lenton, E. A., Gelsthorp, C. H. & Harper, R. (1988). Measurement of progesterone in saliva: assessment of the normal fertile range using spontaneous conception cycles. *Clinical Endocrinology*, **38**, 637–46.

Lenton, E. A., Landgren, B.-M., Sexton, L. & Harper, R. (1984). Normal variation in the length of the follicular phase of the menstrual cycle: effect of chronological age. *British Journal of Obstetrics and Gynaecology*, **91**, 681–4.

Levran, D., Ben-Shlomo, I., Dor, J., Ben-Rafael, Z., Nebel, L. & Mashiach, S. (1991). Age of endometrium and oocytes: observations on conception and abortion rates in an egg donation model. *Fertility and Sterility*, **56**, 1091–4.

Levran, D., Dor, J., Rudak, E., Nebel, L., Ben-Shlomo, I. & Mashiach, S. (1990). Pregnancy potential of human oöcytes – the effect of cryopreservation. *New England Journal of Medicine*, **323**, 1153–6.

Lipson, S. F. & Ellison, P. T. (1992). Normative study of age variation in salivary progesterone profiles. *Journal of Biosocial Science*, **24**, 233–44.

Lipson, S. F. & Ellison, P. T. (1994). Reference values for luteal progesterone measured by salivary radioimmunoassay. *Fertility and Sterility*, in press.

Liu, H. C., Jones, G. S., Jones, H. W. Jr & Rosenwaks, Z. (1988). Mechanisms and factors of early pregnancy wastage in *in vitro* fertilization-embryo transfer patients. *Fertility and Sterility*, **50**, 95–101.

Maslar, I. A. (1988). The progestational endometrium. *Seminars in Reproductive Endocrinology*, **6**, 115–28.

McClure, N., McDonald, J., Kovacs, G. T., Healy, D. L., McCloud, P. I., McQuinn, B. & Burger, H. G. (1993). Age and follicular phase estradiol are better predictors of pregnancy outcome than luteinizing hormone in menotropin ovulation induction for anovulatory polycystic ovarian syndrome. *Fertility and Sterility*, **59**, 729–33.

McNeely, M. J. & Soules, M. R. (1988). The diagnosis of luteal phase deficiency: a critical review. *Fertility and Sterility*, **50**, 1–15.

Meldrum, D. R. (1993). Female reproductive aging – ovarian and uterine factors. *Fertility and Sterility*, **59**, 1–5.

Menken, J., Trussell, J. & Larsen, U. (1986). Age and infertility. *Science*, **233**, 1389–94.

Metcalf, M. G., Donald, R. A. & Livesey, J. H. (1981). Classification of menstrual cycles in pre- and perimenopausal women. *Journal of Endocrinology*, **91**, 1–10.

Metcalf, M. G., Skidmore, D. S., Lowry, G. F. & Mackenzie, J. A. (1983). Incidence of ovulation in the years after menarche. *Journal of Endocrinology*, **97**, 213–19.

Nath, D. C., Singh, K. K., Land, K. C. & Talukdar, P. K. (1994). Breastfeeding and postpartum amenorrhea in a traditional society: a hazards model analysis.

Social Biology, **40**, 74–86.

Navot, D., Bergh, P. A., Williams, M., Garrisi, G. J., Guzman, I., Sandler, B., Fox, J., Schreiner-Engel, P., Hafmann, G. E. & Grunfield, L. (1991). An insight into early reproductive processes through the *in-vivo* model of ovum donation. *Journal of Endocrinology and Metabolism*, **72**, 408–14.

O'Rourke, M. T. & Ellison, P. T. (1993*a*). Salivary estradiol levels decrease with age in healthy, regularly-cycling women. *Endocrine Journal*, **1**, 487–94.

O'Rourke, M. T. & Ellison, P. T. (1993*b*). Age and prognosis in premenopausal breast cancer. *Lancet*, **342**, 60.

Padilla, S. L. & Garcia, J. E. (1989). Effect of maternal age and number of *in vitro* fertilization procedures on pregnancy outcome. *Fertility and Sterility*, **52**, 270–3.

Panter-Brick, C., Lotstein, D. S. & Ellison, P. T. (1993). Seasonality of reproductive function and weight loss in rural Nepali women. *Human Reproduction*, **8**, 684–90.

Peacock, N. R. (1990). Comparative and cross-cultural approaches to the study of human female reproductive failure. In, *Primate Life History and Evolution*, ed. C. J. DeRousseau, pp. 195–220. New York: Wiley-Liss.

Peacock, N. (1991). An evolutionary perspective on the patterning of maternal investment in pregnancy. *Human Nature*, **2**, 351–85.

Pearlstone, A. C., Oei, M. L. & Wu, T.-C. J. (1992). The predictive value of a single, early human chorionic gonadotropin measurement and the influence of maternal age on pregnancy outcome in an infertile population. *Fertility and Sterility*, **57**, 302–4.

Phillips, J., Worthman, C. M. & Stallings, J. F. (1991). New field techniques for detection of female reproductive status. *American Journal of Physical Anthropology, Supplement*, **12**, 143.

Prentice, A. M., Lunn, P. G., Watkinson, M. & Whitehead, R. G. (1983). Dietary supplementation of lactating Gambian women. II. Effect on maternal health, nutritional status and biochemistry. *Human Nutrition and Clinical Nutrition*, **37**, 65–74.

Prentice, A. M. & Prentice, A. (1990). Maternal energy requirements to support lactation. In *Breastfeeding, Nutrition, Infection and Infant Growth and Emerging Countries*, ed. S. A. Atkinson, L. A. Hanson and R. K. Chandra, pp. 67–86. St. John's, Newfoundland, Canada: ARTS Biomedical.

Prentice, A. M. & Whitehead, R. G. (1987). The energetics of human reproduction. *Symposia of the Zoological Society of London*, **57**, 275–304.

Roff, D. A. (1992). *The Evolution of Life Histories*. New York: Routledge, Chapman and Hall.

Rotsztejn, D. A. & Asch, R. H. (1991). Effect of aging on assisted reproductive technologies (ART): experience from egg donation. *Seminars in Reproductive Endocrinology*, **9**, 272–9.

Ruzicka, L. T. & Bhatia, S. (1982). Coital frequency and sexual abstinence in rural Bangladesh. *Journal of Biosocial Science*, **14**, 397–420.

Sauer, M. V., Paulson, R. J. & Lobo, R. A. (1990). A preliminary report on oocyte donation extending reproductive potential to women over 40. *New England Journal of Medicine*, **323**, 1157–60.

Sauer, M. V., Paulson, R. J. & Lobo, R. A. (1992). Reversing the natural decline in human fertility: an extended clinical trial of oocyte donation to women of advanced reproductive age. *Journal of the American Medical Association*, **268**, 1275–9.

Sauer, M. V., Paulson, R. J. & Lobo, R. A. (1993). Pregnancy after 50: application of oocyte donation to women after natural menopause. *Lancet*, **341**, 321–3.

Sherman, B. M. & Korenman, S. G. (1975). Hormonal characteristics of the human menstrual cycle throughout reproductive life. *Journal of Clinical Investigation*, **55**, 699–706.

Sherman, B. M., West, J. H. & Korenman, S. G. (1976). The menopausal transition: analysis of LH, FSH, estradiol, and progesterone concentrations during menstrual cycles of older women. *Journal of Endocrinology and Metabolism*, **42**, 629–36.

Shoham, Z., DiCarlo, C., Patel, A., Conway, G. S. & Jacobs, H. S. (1991). Is it possible to run a successful ovulation induction program based solely on ultrasound monitoring? The importance of endometrial measurements. *Fertility and Sterility*, **56**, 836–41.

Stouffer, R. L. (1988). Perspectives on the corpus luteum of the menstrual cycle and early pregnancy. *Seminars in Reproductive Endocrinology*, **6**, 103–13.

Toner, J. P., Philiput, C. B., Jones, G. S. & Muasher, S. J. (1991). Basal follicle-stimulating hormone level is a better predictor of *in vitro* fertilization performance than age. *Fertility and Sterility*, **55**, 784–91.

Tracer, D. P. (1991). Fertility-related changes in maternal body composition among the Au of Papua New Guinea. *American Journal of Physical Anthropology*, **85**, 393–405.

Treloar, A. E., Boynton, R. E., Behn, B. G. & Brown, B. W. (1967). Variation of the human menstrual cycle through reproductive life. *International Journal of Fertility*, **12**, 77–126.

Udry, J. R., Deven, F. R. & Coleman, S. J. (1982). A cross-national comparison of the relative influence of male and female age on the frequency of intercourse. *Journal of Biosocial Science*, **14**, 1–6.

Venturoli, S., Porcu, E., Fabbri, R., Magrini, O., Paradisi, R., Pallotti, G., Gammi, L. & Famigni, C. (1987). Postmenarchal evolution of endocrine pattern and ovarian aspects in adolescents with menstrual irregularities. *Fertility and Sterility*, **48**, 78–85.

Vihko, R. & Apter, D. (1984). Endocrine characteristics of adolescent menstrual cycles: impacts of early menarche. *Journal of Steroid Biochemistry*, **20**, 231–6.

Virro, M. R. & Shewchuk, A. B. (1984). Pregnancy outcome in 242 conceptions after artificial insemination with donor sperm and effects of maternal age on the prognosis for successful pregnancy. *American Journal of Obstetrics and Gynecology*, **148**, 518–24.

Vitzthum, V. J. (1990). An adaptational model of ovarian function. Res. Rep. no. 90–200, Population Studies Center, University of Michigan, Ann Arbor, MI.

Vitzthum, V. J., Ellison, P. T. & Sukalich, S. In prep. Ovarian function in high altitude Quechua women.

Weinstein, M., Wood, J. & Greenfield, D. D. (1993). How does variation in fetal loss affect the distribution of waiting times to conception? *Social Biology*, **40**, 106–30.

Weinstein, M., Wood, J., Stoto, M. A. & Greenfield, D. D. (1990). Components of age-specific fecundability. *Population Studies,* **44**, 447–67.

Wilson, C., Oeppen, J. & Pardoe, M. (1988). That is natural fertility? The modeling of a concept. *Population Index,* **54**, 4–20.

Wood, J. W. (1989). Fecundity and natural fertility in humans. *Oxford Reviews of Reproductive Biology,* **11**, 61–109.

Wood, J. W. (1990). Fertility in anthropological populations. *Annual Review of Anthropology,* **19**, 211–42.

Wood, J. W., Johnson, P. & Campbell, K. L. (1985). Demographic and endocrinological aspects of low natural fertility in highland New Guinea. *Journal of Biosocial Science,* **17**, 57–79.

Wood, J. W. & Weinstein, M. (1988). A model of age-specific fecundability. *Population Studies,* **42**, 85–113.

Yoshimura, Y. & Wallach, E. E. (1987). Studies of the mechanism(s) of mammalian ovulation. *Fertility and Sterility,* **47**, 22–34.

6 Non-pathological source of variability in fertility: between/ within subjects and between populations

L. ROSETTA

Introduction

For a long time the approach of biological anthropology has been restricted to observational surveys based on questionnaires with a con- comitant limitation to the measurement of variables like the mean age at puberty or the mean duration of post-partum amenorrhea. The quality of data collected by retrospective and/or cross-sectional surveys has been improved with prospective surveys and large sample size. The use of more sophisticated statistical analysis based on probabilities applied to the measurement of the mean age at menarche has given a better estimation of the mean and standard deviation; the application of survival analysis to the estimation of the duration of post-partum amenorrhea among well-defined groups of women has also given more accurate results for this time- dependent variable. The large number of surveys undertaken worldwide has shown the large variability of such variables between populations.

The secular trend in the decrease of the mean age at puberty correlates with better socio-economic conditions, better nutrition and may be less physical activity. There is an apparent stabilization of the phenomenon in industrialized countries, in which earlier puberty has been accompanied by a later mean age at menopause, resulting in longer mean duration of the reproductive life among well-nourished and sedentary populations. In developing countries, where children are still traditionally breastfed for a longer period, recent investigations have shown a trend to shorter durations of post-partum amenorrhea in non-contracepting populations (WHO, unpublished data). A significant improvement in the biological anthropological approach has been the application of hormonal assays in fertility studies, leading to a major distinction with the demographic approach. Plasma assays, difficult to carry out under field conditions, were

91

hopefully challenged by salivary and urinary assays in the last few years. Reference values in hormonal levels have been established under clinical conditions usually in plasma samples and mostly among Western populations. Recently, clinicians have also tested the validity of urinary or salivary assays for easier use even in hospital (Lenton, Gelsthorp & Harper, 1988). The first data collected by biological anthropologist in very different settings, and sometimes very remote areas were blood samples. The results were considered puzzling as compared to the known values for Western subjects (Van der Walt, Wilmsen & Jenkins, 1978). More recently, others have contributed to widen the number of questions arising from such results; the question of normality of the hormonal profile, the range of variability among human beings, or the measurable effect of seasonality on fecundability or early pregnancy loss are all under discussion (Bailey *et al.*, 1992; Leslie, Campbell & Little, 1993). This chapter will review the different aspect of variability in fertility among humans and discuss the role of biological anthropologists in helping to answer these questions.

Intra-individual variability in fertility

Intra-individual variability in human fertility can be related to internal or external causes. The first category is linked to ageing whilst the second is influenced by environmental factors. The role of age in the intra-individual variability in human fertility is certainly the most important source of variability, in both women and men (Menken, Trussell & Larsen, 1986).

The largest dataset collected which quantitatively defines the rhythmic pattern of the human menstrual cycle is that of Treloar *et al.* (1967) which refers to 2700 women collected over a 30-year period. The main characteristics of the menstrual cycle in healthy women can be divided into three different periods. The early post-menarche years showing a large variability of menstrual interval, the period of full reproductive life characterized by a minimum range of variability and the premenopausal years, showing the return to a pattern of mixed short and long intervals. Among regularly menstruating women, a median standard deviation of 1.83 days at age 36 has to be compared to 2.75 days at age 20, but the authors noted the individual characteristics of the human menstrual interval 'each women has her own central trend and variation, both of which change with age' (Treloar *et al.*, 1967).

The mean duration of the menstrual cycle masks more subtle variation in the respective duration of the follicular and the luteal phases, themselves being the result of the hormonal environment at each age. Adolescent cycles compared to normally menstruating adult women usually show a trend towards a longer follicular phase. Immediately after menarche the menstrual pattern seems to be unpredictable in terms of duration and ovulation (Borsos *et al.*, 1986). There is a period of transition during which the secretion of FSH is less than during adulthood with a late selection and a smaller size of the dominant follicle; lower LH surge concentrations and lower luteal phase progesterone concentrations are found in the ovulatory cycles of the adolescents compared to adults. The concept supported by these observations is that, as the sensitivity of the hypothalamic-pituitary axis to the negative feedback effect of gonadal steroids gradually decreases, increasing concentrations of FSH are produced, causing accelerated follicular maturation (Apter *et al.*, 1987). At the other end of reproductive life, in premenopausal women, the change in cycle regularity and cycle length is characterized by shortened follicular phase in the perimenopausal period, normal luteal phase and similar progesterone level as in younger women.

Changes in FSH secretion may be one of the earliest symptoms of human reproductive ageing. It seems that large fluctuations in FSH levels precede the first appearance of gonadal failure. It is still not clear if they can be consecutive either to a decrease in inhibin secretion or to a lower GnRH pulse frequency. Actually, very little information is available concerning changing hypothalamic function with age in women. Irregular menstrual cycles associated with an unpredictable succession of fluctuations are characteristic of the perimenopausal period. They might be accompanied occasionally by abnormal steroid hormone/gonadotrophin ratios, and it is still not clear if there is a change in mean oestradiol concentrations during the perimenopausal and early postmenopausal period (Wise *et al.*, 1991).

Recent data on follicular depletion during menopausal transition have shown a dramatic exponential decrease in primordial follicle numbers during the last ten years of menstruation correlated with the menstrual status from an apparently normal menstrual cycle to irregular periods. The postmenopausal status seems to be associated with total exhaustion of the primordial follicular reserve in women (Richardson, Senikas & Nelson, 1987).

With technically assisted procreation, cross-sectional studies have pointed out that the decrease in fertility starts half-way through in a woman's reproductive life, immediately after 30 years of age. Schwartz and coworkers have studied this aspect in a sample of 2193 women who had azoospermic husbands and were treated by artificial insemination with frozen donor semen (AID) between 1973 and 1980 (Federation CECOS, Schwartz & Mayaux, 1982). The cumulative success rate among women 25 years or less was similar to that for the group 26–30. The decrease in the group 31–35 was significant and even greater in women over 35. Similar decreases in fecundability (conception rate per cycle) with age was shown in the same study. More recently, 72 000 assisted reproductive technology cycles were recorded between 1986 to 1990 in France, representing more than 80% of all French technically assisted procreational activities during this period. The same trend occurred with an additional dramatic decrease in pregnancy rates after 37 years of age (FIVNAT, 1993). The latter finding has been confirmed in other countries: in the USA the pregnancy rate for women undergoing ovulation induction after 40 years of age fell to 3.5% as compared to 13.6% for women younger than 30 years of age (Pearlstone et al., 1992). In Israel, in the group of women participating in an in vitro fertilization (IVF) programme for ovarian failure between 1985 and 1990, those who conceived were significantly younger (median age 31 years) than those who did not (median 37 years) (Levran et al., 1991). In addition among those who conceived the rate of spontaneous abortion was significantly higher for older donors. In that country, egg donation can use only oocytes obtained from women who are themselves undergoing IVF, this means that oocyte from one donor could be distributed to more than one recipient, and fertilized by the recipient's husband's semen. Given that the hormonal therapy was the same for all women, that endogenous ovarian steroid production was negligible in all of them, that the age of oocyte donor was not different between those who conceived and those who did not, and that no male factors were involved, the authors concluded that the aging of the endometrium is a major determinant of reduced fecundity with age as far as implantation is concerned. After implantation of the blastocyst, it is the age of the donor and not the age of the recipient which determines the risk of abortion (mainly due to chromosomal abnormalities) and this suggests that the increasing abortion rate with age should be attributed mainly to aging of the oocyte rather than to the endometrium. Others have recently confirmed that 'a decrease in endometrial receptivity with age is responsible for a higher rate of implantation failure in older women' whilst in patient undergoing egg donation (from young donors) the rate of miscarriage was not different (Yaron et al., 1993).

Hormonal profiles in natural conception cycles ending in early pregnancy loss were compared to successful conception cycles in the same woman (Baird et al., 1991). Urinary oestrone-3-glucuronide and pregnanediol-3-glucuronide were similar up to the time of normal implantation. This result suggests that most early losses in reproductively normal women do not result directly from deficiencies in ovarian steroid production but from a possible delay in implantation, presumably due to a delay in the chorionic gonadotrophin (hCG) signal which came too late to effectively maintain the corpus luteum function. This could be due to late fertilization, slow tubal transport, slow development of the conceptus after fertilization, or problems with the early stages of the implantation process.

In men, even if there is no clear limit in the ability to conceive, there is an effect of age on semen characteristics. From 1973 to 1980, a study was carried out on more than 800 fertile men aged between 20 and 50, having already fathered at least one child, and belonging either to a group of prevasectomy subjects who were requesting semen cryopreservation prior to sterilization, or to semen donor candidates at the Centre d'Etude et de Conservation du Sperme Humain (CECOS), Paris-Bicêtre in France (Schwartz *et al.*, 1983). The 20–25 year old age group has lower values than within the 26–30 age group or 31–35 age group for almost all qualitative characteristics, whilst the values for older subjects were significantly lower for at least one characteristic: prefreeze motility was lower in the 46 to 50 year age group. In addition, the morphological features in the group 41–45 year age group and postthaw motility in the group 36–40 year age group were significantly lower than in younger age groups. The finding that sperm production does decline significantly with age has been confirmed by Johnson, Petty and Neaves (1984). The decrease in daily sperm production continues to decline after 50 years of age. It was estimated as approximately 30% less in the age group 50–80 compared to the group of 21–50 years old. The male–female age effects can be compared; there is an improvement in semen quality up to 25 years of age, followed by a plateau corresponding to maximal fertility between 25 and 35, then from 35 onwards appears the first signs of decrease in fertility parameters in both genders.

For men and women, the external factors relating to lifestyle which may have an effect on individual variability in fertility have been reviewed (Rosetta, 1993*a*). They are linked to nutrition and/or high level of endurance physical activity. It has been shown that voluntary healthy women with a normal menstrual cycle undergo a modification in the duration of the follicular phase after diet manipulation during a complete previous menstrual cycle (Hill *et al.*, 1984; Longcope *et al.*, 1987; Cassidy, Bingham & Setchell, 1994). The effect of environment, like seasonality in terms of photoperiod or temperature, is not likely to reach a significant effect in human female fecundability, except in very contrasted climates like subarctic region where an increase in either, endometrium receptivity or a better quality of the ova, seems to correlate with luminosity more than temperature (Rönnberg *et al.*, 1990) but this finding is not acknowledged in most of other countries (Mayaux & Spira, 1989; Rosetta, 1993*b*). Nevertheless, in men, a significant reduction in sperm concentration, total sperm count per ejaculates and motile sperm has been shown during the summer in healthy American men working outdoors most of the time in warm climates (Levine *et al.*, 1990). Among several possible confounding factors analysed, season was the only significant predictor of the variability in concentration-related measures of semen quality within subjects.

Inter-individual variability in fertility

There is a large inter-individual variability in fertility from puberty to menopause in women but also from maturation to adulthood in men. At puberty, there is a large continuum from early maturer to late maturer and this is mirrored all along the reproductive life in terms of fully successful fertile subjects to less successful, and finally infertile men or women, with possible shift from one status to another one when this is due to external factors like physical workloads or high level of physical training or dietary factors.

Women with early menarche developed a regular cycle pattern more promptly than women with late menarche: a comparison of early maturer, medium and late maturer, respectively, has shown that 50% of the menstrual cycles were ovulatory at about 1, 3, and 4.5 years after menarche in each group respectively (Apter & Vihko, 1983).

In adult women, there is variation in the duration of menstrual flow, the follicular phase and luteal phase and in the duration of the rise in body temperature associated with ovulation. As if the delay observed during the follicular phase was an indicator of a less successful reproductive power, longer duration of menstrual period, longer duration of the hypothermic phase, longer duration of temperature rise at the time of ovulation are all associated with higher incidence of congenital malformations (Spira *et al.*, 1985). In addition, the general reproductive characteristics of women having given birth to malformed infants were those with later age at menarche and higher rate of irregular periods. The risk of congenital malformation was 9.5% when the hypothermic phase lasted 24 days or more, versus 1.3% for shorter length of hypothermic phase.

Irregular menses generally reflect hypoestrogenic conditions. It is clear from many studies that oestrogen concentrations are significantly lower in postmenopausal women who have been acyclic for 5–10 years. But there is also evidence of menstrual differences associated with vegetarian and non-vegetarian diet: higher incidence of menstrual irregularities has been found in traditional vegetarian compared to non-vegetarian pre-menopausal healthy women (Pedersen *et al.*, 1991). Beside individual factors, mainly inherited, environmental factors are probably able to influence the inter-individual variability in fertility.

Another group of data useful to evaluate inter-individual variability in fertility are those related to the return of fertility after pregnancy, either in breastfeeding or non-breastfeeding women. Brown, Harrison and Smith (1985) have published the results of longitudinal studies on individuals and shown a large inter-individual variability. This has been confirmed recently in another study showing that the resumption of menses and ovulation in a group of well-nourished breast-feeding women had a wide range; the earliest ovulation and menstruation recorded was 24 days and 35 days, respectively, whilst in the same study the latest ovulation was at 750 days and the latest menstruation at 698 days (Lewis *et al.*, 1991). The 24 h pattern of pulsatile luteinizing hormone (LH), follicle stimulating hormone (FSH) and prolactin in fully breast-feeding women at 4 and 8 weeks post-partum has also shown the intra-individual variability in the resumption of ovarian activity, some subjects resuming LH pulses earlier than others; the frequency also was found to vary considerably between subjects (Tay, Glasier & McNeilly, 1992). Assuming that each LH pulse was induced by the resumption of the pulsatility of the gonadotropin-releasing hormone from the hypothalamus, this study clearly shows the variability in sensitivity of the hypothalamo-pituitary axis after pregnancy. In addition, early differences in the endocrine profile of long and short lactational amenorrheic women have been shown in Chilean breast-feeding women, those with short duration of post-partum amenorrhoea having higher E2 levels than their counterparts with longer amenorrhoea (Diaz *et al.*, 1991).

Inter-individual variability in fertility in men

Variation amongst individual men has been reported in daily sperm production and daily sperm output in ejaculates (Johnson *et al.*, 1984). In addition, significantly lower sperm characteristics has been found in brothers of infertile men compared with fertile controls (Czyglik *et al.*, 1986) raising the causal effect of either genetic or environmental factors. The population studied was composed of men who volunteered to become semen donors for artificial insemination. They all provided a semen sample used to select acceptable donors. All of them had already had at least one child, some of them were brothers of husbands in couples awaiting artificial insemination with donor, implying that the husband was either azoospermic or severely oligozoospermic. All the semen samples were obtained according to the same protocol and the semen characteristics of the brothers of infertile men ($n = 36$) were compared with the group of non-related donors as controls ($n = 545$). They show significant lower characteristics for sperm count, motility and morphology, but not volume that the controls. The related samples were divided according to the characteristics of husband infertility and compared to controls subjects: brothers of azoospermic men ($n = 20$) had significantly lower sperm count than brothers of oligozoospermic men ($n = 12$) even if it still remained in normal limits.

Variability in fertility between populations

As there is individuals more fertile than others, it seems that there is a gradient in fertility between populations, at least between the non-agricultural and agricultural populations (Bentley *et al.*, 1993). There is some evidence of an increase in fertility with better socio-economic level, better nutrition and lower level of physical activity, and/or higher level of energy balance. Apart from pathological causes, a trend in higher natural fertility must be expected from hunter–gatherers, then pastoralists and poor traditional farmers without mechanization, to very poor urbanized populations living in slums areas. External factors like nutrition and physical activity are likely to play a significant role in the regulation of fertility only in very contrasted situations (Goodman *et al.*, 1985). In this context, sedentarization and urbanization might be correlated with different levels of natural fertility (Rosetta, 1995). Several other causes have been suspected of playing a role in subfertility, or infertility. Infectious diseases has been identified as a major cause of lower fertility in populations without easy access to medical care (Pennington & Harpending, 1991; Jenkins, 1993). A chapter is dedicated to this topic in this book by Mascie-Taylor. Others have clearly pointed out the role of breastfeeding pattern in the change of duration in post-partum amenorrhea among non-contracepting populations (Wood, Johnson & Campbell, 1985; Liestøl, Rosenberg & Walløe, 1988; Vitzthum, 1989), but different aspects of the variability in fertility between populations has been investigated with the introduction of hormonal assays in free living conditions.

A survey was set up to test the dietary influence on the oestrogen secretion in a population of hunter–gatherers in Botswana compared to Black African controls (Van der Walt *et al.*, 1978). The mean hormonal profile for men was in the normal range, whilst a significantly lower level of oestrogen was shown for the whole sample of women of child-bearing age compared to healthy Black African women of the same age. Only one of the female hunter–gatherers tested had a level of progesterone compatible with an ovulatory cycle but the level of gonatrophin was in the normal range. The cause of this ovarian suppression was not identified but several hypothesis has been proposed, either a direct effect of the diet on the oestrogen metabolism, or a difference in clearance of oestrogen metabolites, or variation in the mean hormonal profile between populations.

Some years later in an attempt to identify risk factors of breast cancer in women from different cultures, investigations were carried out to test the influence of the diet on oestrogen secretion and excretion. The comparison in Hawaii, between Oriental migrant women and American omnivorous women has shown that a vegetarian diet was associated with significantly lower incidence of breast cancer and significantly lower oestrogen levels (Goldin *et al.*, 1986). The difference in food composition was a lower fat content, less calorie intake and higher fibre intake among the Oriental women eating a mainly vegetarian diet. The level of plasma oestrone and a different pattern of urinary to faecal excretion between the two groups were positively correlated to the amount of total fat and saturated fat intake.

Other epidemiological surveys carried out in large sample of various populations for the same purpose have confirmed this finding. Plasma concentration of oestradiol suspected to modify the incidence of breast cancer were significantly lower among Chinese women compared to British controls in the same age group and the same reproductive status. This survey carried out in China has involved 3250 Chinese women between 35 and 64 years of age in 65 counties and 100 British controls in each of the 3 age classes (35–44, 45–54, 55–64). In both populations, mean oestradiol concentration decreased with age, the difference is constantly lower for Chinese women whatever the age group (Key *et al.*, 1990).

Others have used progesterone assays to investigate the ovarian function in different groups of women. Daily measurement of salivary or urinary progesterone during a whole menstrual cycle gives several indications about the possible impairment of fertility or fecundability in women. A normal ovulation is accompanied by a rise in progesterone at the time of ovulation, a delay in the surge of progesterone or a short luteal phase are all indicators of subfertility. In addition, it has been shown that the mean progesterone level during the plateau of the luteal phase may vary and a lower level is usually associated with lower fertility. Reference values for progesterone concentration in saliva in normal fertile women living in the United Kingdom has been established for cycles in which conception has occurred and compared to normal ovulatory cycles (Lenton *et al.*, 1988). The geometric mean concentrations of progesterone were lower and the variability was wider in the non-conceptional cycles. Using salivary progesterone, Ellison and co-workers have shown a high incidence of anovulatory cycles among African women from the Ituri Forest, Zaire compared to American controls, and for those with an ovulatory cycle the progesterone profile was significantly lower than those established for the American controls (Ellison, Peacock & Lager, 1989).

During the post-partum period the return of luteinization was compared between three groups of nursing women, one from rural Zaire, another one from urban Zaire, and a group of Swedish women (Hennart *et al.*, 1985). The return of luteinization was significantly delayed in both groups of women from Zaire.

Whatever the cause, there is evidence of variability in female fertility between populations. The duration of the menstrual cycle was 40% longer among women of Highland New Guinea, than the median estimated for US controls matched for gynaecological age (Johnson *et al.*, 1987). Higher rate of anovulation in traditional populations, frequently inadequate luteal phases, lower mean hormonal steroid level during ovulatory cycles among vegetarian compared to omnivorous, or a trend in earlier maturation for Western adolescent girls compared to schoolgirls from the same socio-economic level in Asian country (Danutra *et al.*, 1989), are all indicators for lower fertility among a large proportion of women living in developing countries.

Very little has been done to investigate the variability in fertility in males. The comparison of testosterone levels among men living at high altitude has confirmed the trend in lower values among older high-altitude native Bolivian compared to young men in the same population but the testosterone values for both groups were in the same range as values found at sea level (Beall *et al.*, 1992), whilst a seasonal decrease in testosterone values among Lese men living in the Ituri Forest, Zaire has been shown (Bentley, Goldberg & Jasieñska, 1993).

Another fascinating application for hormonal assays, to widen our knowledge about the regulation of fertility, is the detection of pregnancy in the early stages and the estimation of fetal loss in free-living populations. Leslie *et al.* (1993) have already published the results of prospective surveys carried out among a sample of nomadic and settled Turkana women involved in the South Turkana Ecosystem Project between April 1989 and August 1990 (Little, Leslie & Campbell, 1993). Each of the 335 women provided early morning urinary samples for 3 consecutive days, in which chorionic gonadrophin (hCG), luteinizing hormone (LH) and pregnandiol glucuronide (PdG) were determined by solid-phase enzyme immuno-assays in the field. The sensitivity of the technique was estimated to allow the detection of pregnancy as soon as one week after implantation. Body size and body composition were estimated by anthropometric measurements and a recall of reproductive history, recent menstrual function, diet and health was done concomitantly. At each of the two surveys in each group, the proportion of women pregnant was about twice in the settled women (24 and 19%) than in the nomadic (10 and 13%), but none of the detected and followed pregnancy among the nomadic was lost, whilst 45% among the settled women were lost, most of them during the first trimester. This puzzling result have not been fully explained by the authors, but it highlights the utility of hormonal assays in the biological anthropological approach of fertility.

Conclusions

This review has highlighted knowledge, progress and limitations of our understanding about the variability in human fertility. One of the questions arising over the last 15 years was about the normality of hormonal values across populations and the range of variability compatible with fertility. Most of the references values have been established under clinical conditions in developed countries, and lower values for gonadal hormone are often shown for women having a different lifestyle and different diet. Does it mean that the range of variability established for mostly Western women are applicable to other populations. Are different mean levels of oestrogen and progesterone compatible with fecundability and the maintenance of a normal pregnancy under different conditions of nutrition or physical activity?

Hormonal assays are also a tool to understanding seasonal variation of fertility in the same population. In warm climates where the variation in temperature are likely to influence sperm quality it is of importance to check for female fertility variables like the rate of anovulation or the possible modification in hormonal profile during seasonal stress.

To answer all these questions, several field surveys with hormonal assessment among representative sample will be necessary to compare populations and to establish local standards if necessary. We are certainly at the beginning of an enthusiastic effort in this direction from biological anthropologists involved in the study of fertility. If it is confirmed that the high level of mean hormonal steroid usually found among Western women is associated with higher risk of hormonal-dependent cancer, it can no longer be seen as a desirable standard for every women worldwide, and there is a place for biological anthropology to help in the findings about the variability in fertility.

REFERENCES

Apter, D. & Vihko, R. (1983). Early menarche, a risk factor for breast cancer, indicates early onset of ovulatory cycles. *Journal of Clinical Endocrinology and Metabolism*, **57**, 82–6.

Apter, D., Räisänen, I., Ylöslato, P. & Vihko, R. (1987). Follicular growth in relation to serum hormonal patterns in adolescent compared with adult menstrual cycles. *Fertility and Sterility*, **47**, 82–8.

Bailey, R. C., Jenike, M. R., Ellison, P. T., Bentley, G. R., Harrigan, A. M. & Peacock, N. R. (1992). The ecology of birth seasonality among agriculturalists in central Africa. *Journal of Biosocial Science*, **24**, 393–412.

Baird, D. D., Weinberg, C. R., Wilcox, A. J., McConnaughey, D. R., Musey, P. I. & Collins, D. C. (1991). Hormonal profiles of natural conception cycles ending in early, unrecognized pregnancy loss. *Journal of Clinical Endocrinology and Metabolism*, **72**, 793–800.

Beall, C. M., Worthman, C. M., Stallings, J., Strohl, K. P., Brittenham, G. M. & Barragan, M. (1992). Salivary testosterone concentration of Aymara men native to 3600 m. *Annals of Human Biology*, **19**, 67–78.

Bentley, G. R., Goldberg, T. & Jasieñska, G. (1993). The fertility of agricultural and non-agricultural traditional societies. *Population Studies*, **47**, 269–81.

Bentley, G. R., Harrigan, A. M., Campbell, B. & Ellison, P. T. (1993). Seasonal effects on salivary testosterone levels among Lese males of the Ituri forest, Zaire. *American Journal of Human Biology*, **5**, 711–17.

Borsos, A., Lampe, L. G., Balogh, A., Csoknyay, J. & Ditroi, F. (1986). Ovarian function immediately after the menarche. *International Journal of Gynaecology and Obstetrics*, **24**, 239–42.

Brown, J. B., Harrison, P. & Smith, M. A. (1985). A study of returning fertility after child-birth and during lactation by measurement of urinary oestrogen and pregnanediol excretion and cervical mucus production. *Journal of Biosocial Science*, Supplement 9, 5–23.

Cassidy, A., Bingham, S. & Setchell, D. R. (1994). Biological effects of a diet of soy protein rich in isoflavones on the menstrual cycle of premenopausal women: implications for the prevention of breast cancer. *American Journal of Clinical Nutrition*, **60**, 333–40.

Czyglik, F., Mayaux, M.-J., Guihard-Moscato, M.-L., David, G. & Schwartz, D. (1986). Lower sperm characteristics in 36 brothers of infertile men, compared with 545 controls. *Fertility and Sterility*, **45**, 255–8.

Danutra, V., Turkes, A., Read, G. F., Wilson, D. W., Griffiths, V., Jones, R. & Griffiths, K. (1989). Progesterone concentrations in samples of saliva from adolescent girls living in Britain and Thailand, two countries where women are at widely different risk of breast cancer. *Journal of Endocrinology*, **121**, 375–81.

Díaz, S., Cárdenas, H., Brandeis, A., Miranda, P., Schiappacasse, V., Salvatierra, A. M., Herreros, C. Serón-Ferré, M. & Croxatto, H. B. (1991). Early difference in the endocrine profile of long and short lactational amenorrhea. *Journal of Clinical Endocrinology and Metabolism*, **72**, 196–201.

Ellison, P. T., Peacock, N. R. & Lager, C. (1989). Ecology and ovarian function among Lese women of the Ituri forest, Zaire. *American Journal of Physical Anthropology*, **78**, 519–26.

Federation CECOS, Schwartz, D. & Mayaux, M. J. (1982). Female fecundity as a function of age. Results of artificial insemination in 2193 nulliparous women with azoospermic husbands. *New England Journal of Medicine*, **306**, 404–6.

FIVNAT (French *In Vitro* National) (1993). French National IVF registry: analysis of 1986 to 1990 data. *Fertility and Sterility*, **59**, 587–95.

Goldin, B. R., Adlercreutz, H., Gorbach, S. L., Woods, M. N., Dwyer, J. T., Conlon, T., Bohn, E. & Gershoff, S. N. (1986). The relationship between estrogen levels and diets of Caucasian American and Oriental immigrant women. *American Journal of Clinical Nutrition*, **44**, 945–53.

Goodman, M. J., Estioko-Griffin, A., Griffin, P. B. & Grove, J. S. (1985). Menarche, pregnancy, birth spacing and menopause among foragers of Cayagan Province, Luzon, the Philippines. *Annals of Human Biology*, **12**, 169–77.

Hennart, P., Hofvander, Y., Vis, H. & Robyn, C. (1985). Comparative study of nursing mothers in Africa (Zaire) and in Europe (Sweden): breastfeeding behaviour, nutritional status, lactational hyperprolactinaemia and status of the menstrual cycle. *Clinical Endocrinology*, **22**, 179–87.

Hill, P., Garbaczewski, L., Haley, N. & Wynder, E. L. (1984). Diet and follicular development. *American Journal of Clinical Nutrition*, **39**, 771–7.

Jenkins, C. (1993). Fertility and infertility in Papua New Guinea. *American Journal of Human Biology*, **5**, 75–83.

Johnson, L., Petty, C. S. & Neaves, W. B. (1984). Influence of age on sperm production and testicular weights in men. *Journal of Reproduction and Fertility*, **70**, 211–18.

Johnson, P. L., Wood, J. W., Campbell, K. L. & Maslar, I. A. (1987). Long ovarian cycles in women of highland New Guinea. *Human Biology*, **59**, 837–45.

Key, T. J. A., Chen, J., Wang, D. Y., Pike, M. C. & Boreham, J. (1990). Sex hormones in women in rural China and in Britain. *British Journal of Cancer*, **62**, 631–6.

Lenton, E. A., Gelsthorp, C. H. & Harper, R. (1988). Measurement of progesterone in saliva: assessment of the normal fertile range using spontaneous conception cycles. *Clinical Endocrinology*, **38**, 637–46.

Leslie, P. W., Campbell, K. L. & Little, M. A. (1993). Pregnancy loss in nomadic and settled women in Turkana, Kenya: a prospective study. *Human Biology*, **65**, 237–54.

Levine, R. L., Mathew, R. M., Brandon Chenault, C., Brown, M. H., Hurtt, M. E., Bentley, K. S., Mohr, K. L. & Working, P. K. (1990). Differences in the quality of semen in outdoor workers during summer and winter. *The New England Journal of Medicine*, **323**, 12–16.

Levran, D., Ben-slomo, I., Dor, J., Ben-Rafael, Z., Nebel, L. & Mashiach, S. (1991). Aging of endometrium and oocytes: observations on conception and abortion rates in an egg donation model. *Fertility and Sterility*, **56**, 1091–4.

Lewis, P. R., Brown, J. B., Renfree, M. B. & Short, R. V. (1991). The resumption of ovulation and menstruation in a well-nourished population of women breastfeeding for an extended period of time. *Fertility and Sterility*, **55**, 529–36.

Liestøl, K., Rosenberg, M. & Walløe, L. (1988). Lactation and post-partum amenorrhea: a study based on data from three Norwegian cities 1860–1964. *Journal of Biosocial Science*, **20**, 423–34.

Little, M. A., Leslie, P. W. & Campbell, K. L. (1993). Energy reserves and parity of nomadic and settled Turkana women. *American Journal of Human Biology*, **4**, 729–38.

Longcope, C., Gorbach, S., Goldin, B., Woods, M., Dwyer, J., Morrill, A. & Warram, J. (1987). The effect of a low fat diet on estrogen metabolism. *Journal of Clinical Endocrinology and Metabolism*, **64**, 1246–50.

Menken, J., Trussell, J. & Larsen, U. (1986). Age and infertility. *Science*, **233**, 1389–94.

Mayaux, M.-J. & Spira, A. (1989). Seasonal distribution in conceptions achieved by artificial-insemination by donor. Letter. *British Medical Journal*, **298**, 167.

Pearlstone, A. C., Fournet, N., Gambonne, J. C., Pang, S. C. & Buyalos, R. P. (1992). Ovulation induction in women age 40 and older: the importance of basal follicle-stimulating hormone level and chronological age. *Fertility and Sterility*, **58**, 674–9.

Pedersen, A. B., Bartholomew, M. J., Dolence, L. A., Aljadir, L. P., Netteburgh, K. L. & Lloyd, T. (1991). Menstrual differences due to vegetarian and nonvegetarian diets. *American Journal of Clinical Nutrition*, **53**, 879–85.

Pennington, R. & Harpending, H. (1991). Infertility in Herero pastoralists of Southern Africa. *American Journal of Human Biology*, **3**, 135–53.

Richardson, S. J., Senikas, V. & Nelson, J. F. (1987). Follicular depletion during menopausal transition: evidence for accelerated loss and ultimate exhaustion. *Journal of Clinical Endocrinology and Metabolism*, **65**, 1231–7.

Rönnberg, L., Kauppila, A., Leppäluoto, J., Martkainen, H. & Vakkuri, O. (1990). Circadian and seasonal variation in human preovulatory follicular fluid melatonin concentration. *Journal of Clinical Endocrinology and Metabolism*, **71**, 493–6.

Rosetta, L. (1993a). Female reproductive dysfunction and intense physical training. In *Oxford Reviews of Reproductive Biology*, Vol. 15, ed. S. R. Milligan. pp. 113–41. Oxford: Oxford University Press.

Rosetta, L. (1993b). Seasonality and fertility. In *Seasonality and Human Ecology*, SSHB Symposium 35, pp. 65–75, ed. S. J. Ulijaszek & S. S. Strickland, Cambridge: Cambridge University Press.

Rosetta, L. (1995). Nutrition, physical workloads and fertility. In *Human Reproductive Decisions: Biological and Social Perspectives*, ed. R. I. M. Dunbar. pp. 52–75, London: Macmillan.

Schwartz, D., Mayaux, M. J., Spira, A., Moscato, M-L., Jouannet, P., Czyglik, F. & David, G. (1983). Semen characteristics as a function of age in 833 fertile men. *Fertility and Sterility*, **39**, 530–5.

Spira, A., Spira, N., Papiernik-Berkauer, E. & Schwartz, D. (1985). Pattern of menstrual cycles and incidence of congenital malformations. *Early Human Development*, **11**, 317–24.

Tay, C. C. K., Glasier, A. F. & McNeilly, A. S. (1992). The 24 h pattern of pulsatile luteinizing hormone, follicle stimulating hormone and prolactin release during the first 8 weeks of lactational amenorrhoea in breastfeeding women. *Human Reproduction*, **7**, 951–8.

Treloar, A. E., Boynton, R. E., Behn, B. G. & Brown, B. W. (1967). Variation of the human menstrual cycle through reproductive life. *International Journal of Fertility*, **12**, 77–126.

Van der Walt, L. A., Wilmsen, E. N. & Jenkins, T. (1978). Unusual sex hormone patterns among desert-dwelling hunter–gatherers. *Journal of Clinical Endocrinology and Metabolism*, **46**, 658–63.

Vitzthum, V. J. (1989). Nursing behaviour and its relation to duration of post-partum amenorrhea in an Andean community. *Journal of Biosocial Science*, **21**, 145–60.

Wise, P. M., Scarbrough, L., Larson, G. H., Lloyd, J. M., Weiland, N. G. & Chiu, S. F. (1991). Neuroendocrine influences on aging of the female reproductive system. *Frontiers in Neuroendocrinology*, **12**, 323–56.

Wood, J. W., Johnson, P. L. & Campbell, K. L. (1985). Demographic and endocrinological aspects of low natural fertility in Highland New Guinea. *Journal of Biosocial Science*, **17**, 57–79.

Yaron, Y., Botchan, A., Amit, A., Kogozowski, A., Yovel, I. & Lessing, J. B. (1993). Endometrial receptivity – the age related decline in pregnancy rates and the effect of ovarian function. *Fertility and Sterility*, **60**, 314–18.

7 The relationship between disease and subfecundity

C. G. N. MASCIE-TAYLOR

Introduction

There is enormous variation in individual fertility. *The Guinness Book of Records 1993* (1992) reports the greatest officially recorded number of children born to one mother is 69, by the wife of a Russian peasant. In 27 confinements she gave birth to 16 pairs of twins, seven sets of triplets and four sets of quadruplets. The case was reported to Moscow by the Monastery of Nikolskiy on 27 February 1782. Currently, the most prolific mother is a Chilean, who in 1981, produced her 55th and last child. In polygamous societies the number of descendants can become incalculable. The last Sharifian Emperor of Morocco, Moulay Ismail (1672–1727) known as 'The Bloodthirsty' was reputed to have fathered a total of 525 sons and 342 daughters by 1703 and achieved a 700th son in 1721.

McFalls and McFalls (1984) estimate the maximum population fecundity, i.e. the average fecundity of the individual members of a population, at about 15 children per women. On that basis all populations are to a greater, or lesser extent, subfecund. The Hutterites, a religious sect living in North America, are generally acknowledged to be the most fertile population with an average of about 8 children per women while populations in the Bas-Uele district in Zaire illustrate the opposite extreme; nearly 50% of women aged 30–34 were found to be childless in the 1960s (Romaniuk, 1968) compared to only 2% of Hutterite women (Eaton & Mayer, 1954).

There are a number of reasons why the number of children per women varies within and between populations. If we compare the Hutterites and Zaireans, then it is very obvious that the Hutterites have the benefits of good nutrition, access to advanced health care facilities, live in an environment where infectious diseases and environmental hazards are mainly absent and where sexual relations only occur within marriage. In comparison, the population of Bas-Uele suffers from poverty, poor nutrition, inadequate health facilities and where infectious diseases like malaria, are endemic, and where the number of sexual partners varies. In

106

Table 7.1. *Disease and fertility*

1. Delay or prevent entry into marriage
2. Bring about marital disruption leading to divorce, separation or desertion
3. Voluntary abstinence to avoid spreading disease
4. Involuntary abstinence because of hospitalization
5. Increase the use of contraception
– to avoid affecting partner and/or fetus
– to avoid childbearing because of disease
6. Increased likelihood of sterilization
7. Lessening of sexual desire leading to lower coital frequency
8. Induced abortion
9. Subfecundity
– coital inability
– conceptive failure
– pregnancy loss

addition, there will be differences in the numbers of individuals using birth control and coital activity will vary.

Disease and fertility

There are a number of ways in which disease affects fertility and these have been thoroughly reviewed by McFalls and McFalls (1984) and are summarized in Table 7.1. In addition, fertility can be influenced by the known relationship between disease and migration; people are known to leave disease ridden areas and individuals with, for instance, leprosy are often forced to migrate because of their affliction.

    This chapter is not concerned with trying to identify or to describe all the factors which impinge upon differential fertility rates for which there is a considerable literature (see Davis & Blake, 1956; Bongaarts, 1983). Instead, it focuses on population subfecundity.

Prevalence of subfecundity

Developed countries

It is difficult to ascertain exact levels of subfecundity. For instance, in the USA Mosher and Pratt (1982) found that 25.3% of USA couples with wives in the age range 15–44 were subfecund (Table 7.2). Although this figure includes women who were sterilized for non-contraceptive reasons, it

Table 7.2. *Subfecundity in United States women*

		Subfecund (%)					
Age Group	Number of women (in 1000s)	Non-contra-ceptively sterile	Non-surg-ically sterile	Other Problems	Long Birth Interval	Total	Contra-ceptively sterile
15–19	1,043	0.2	0.0	8.8	0.1	9.1	0.8
20–24	4,977	0.5	0.3	10.0	1.0	11.7	4.0
25–29	6,443	4.0	1.3	11.1	2.3	18.7	12.5
30–34	5,736	9.8	1.3	12.0	2.9	26.0	26.4
35–39	4,814	16.4	2.3	9.9	6.2	34.8	28.8
40–44	4,474	22.6	1.8	8.7	9.3	42.4	31.2
Total	27,488	9.6	1.3	10.4	3.9	25.3	18.5

Table 7.3. *Infertility diagnosis rates in France 1986–1990*

	Year				
	1986	1987	1988	1989	1990
Tubal Only	55.2	48.5	48.3	42.0	40.7
Male Only	10.0	11.0	11.2	12.3	13.5
Unknown	9.7	9.8	10.1	14.2	12.0

probably seriously underestimates the level of subfecundity since it cannot detect subfecund or sterile contraceptive users. In addition, the survey excluded the unmarried and those not married for a full 12 months preceding the interview.

Subfecundity can result from either coital inability, infertility and pregnancy loss. Detailed information on coital ability has been obtained by Masters and Johnson (1970) and by Frank, Anderson and Rubinstein (1978) in the USA. Both research teams find that between 40% and 50% of couples experience coital difficulties at sometime during their lives. Infertility or conceptive difficulty is also common and about 10% of married couples are thought to be infertile (Mosher, 1982).

Recent data are available from national or seminational *in vitro* fertilization (IVF) registers. In a French survey (Table 7.3) covering the years 1986 to 1990 (Favnat, 1993), the percentage diagnosed as infertile of tubal origin fell from 55.2% to 40.7% whereas the 'male only' increased throughout this period. The proportion of women presenting a tubal disease remained stable at about two thirds of the patients while infertility of unknown origin remained at 10% of all cases in the 5-year study. Similar

Table 7.4. *Percentages of childlessnesses in selected countries of Africa*

Country	Percentage of childless women	
	25–29 years	50+ years
Cameroon – north	21.0	15.0
– southeast	28.0	23.0
– west	7.0	6.7
Centrally African Republic	25.2	13.6
Gabon	34.0	31.9
Niger	12.8	5.2
Mali	26.0	15.0
Senegal	12.0	5.6
Sudan – overall	—	9.6
– Kassala	—	13.5
Zaire – overall	22.1	17.6
– Bas-Uele	50.7	37.3

percentages for women were found in an American study with tubal disease accounting for 62% and unknown 10.5% (Medical Research International, 1990).

About 30% of married US women in the age range 35–44 are reported as experiencing at least one spontaneous pregnancy loss and each year between 15% and 20% of all conceptions (about 800 000 pregnancies) end as a spontaneous abortion. Thus the data on coital inability, infertility and pregnancy loss all suggest subfecundity is extensive in developed countries.

Developing countries

Turning to developing countries the picture is difficult to assess because of the paucity of reliable data. Consequently, the childlessness rate of women who have completed their reproductive period is used as a proxy for subfecundity. High rates of childlessness have been reported in many parts of Africa including Cameroon, Tanzania, Zaire, Sudan, Central African Republic, Gabon and Chad (Table 7.4). Such reports might seem paradoxical given the high overall African birth rate.

Even within countries regional variation is found: for instance, in Zaire, childlessness in women aged from 45–49 years ranges from a low of 6% among the Batwa-Batshwa and Ngbaka peoples to a high of 65% in the Mbelo; in the three southern states of Sudan the range is from 1 to 42%.

The impact of non-sexually transmitted diseases on subfecundity

Amoebiasis

Amoebiasis is an ulcerative and inflammatory disease of the colon caused by the protozoan *Entamoeba histolytica*. The disease is widespread in tropical countries, and over 40% of the populations may be infected; even in the USA the prevalence rate is about 3–4%. Infection occurs by ingestion of the cyst of *E. histolytica* in food or water contaminated with faeces. Once in the small bowel the cysts rupture releasing 8 trophozoites. The trophozoites migrate to the colon where they go through binary fission every 8 hours. Clinical symptoms range from asymptomatic carrier state to fulminant dysentery.

Several studies have suggested that amoebiasis during pregnancy may be more severe and may be associated with higher mortality rate than occurs in nonpregnant women. For instance, Abioye (1973) found that two-thirds of fatal cases of amoebiasis in females were associated with pregnancy. The reasons for the elevated risk in pregnancy have been discussed by Reinhardt (1980) and they include (a) increased levels of free plasma cortisol during pregnancy (b) increased levels of serum cholesterol in early pregnancy and (c) malnutrition and anaemia which are commonly present in women living in areas where amoebiasis is endemic.

There is no evidence the *E. histolytica* can be transported across the placenta and thus there is no direct impact on the foetus. Even so it has been proposed that amoebiasis during pregnancy may have an adverse effect upon the fetus. Czeizel, Hancsok and Palkowich (1966) found a significantly higher incidence of positive stool cultures of *E. histolytica* in women having spontaneous abortions compared with women having term births. In addition, they found that women who gave birth to infants with congenital anomalies had an increased incidence of amoebiasis. Finally, the newborn child who acquires the disease (usually from the mother) can suffer serious illness with hepatic abscess, gangrene of the colon and colon perforation with peritonitis.

Giardiasis

Giardiasis is a common disease found worldwide particularly in South America and Mexico, Africa (west and central), Asia (southeast and south) and the Soviet Union. The disease is caused by *Giardia lamblia*, a flagellated protozoan parasite which inhabits the duodenum. In the United States analysis of stool samples indicates that *G. lamblia* is the most commonly identified pathogenic intestinal parasite occurring in 3–9% of samples. In its most severe form the disease results in severe diarrhoea with malabsorption.

It is generally agreed that giardiasis has little or no effect on pregnancy outcome. However there is some evidence that significant malabsorption may impair fertility and adversely affect pregnancy as suggested by Kreutner, Del Bene and Amstey's (1981) study of three cases of severe giardiasis.

Tuberculosis

The agent for Tuberculosis (TB) is the bacterium *Mycobaterium tuberculosis*. Although tuberculosis is usually associated with the formation of nobular lesions (tubercules) in the lungs, other organs can be invaded. Genital TB in both males and females can be a cause of painful coitus; in males it leads to seminal vesiculitis, orchitis or chronic prostatis with concomitant dyspareunia or even impotence, while women with the disease suffer from pelvic pain which may be aggravated by coitus. In addition pulmonary TB is likely to lead to reduced libido or lack of sex drive in both males and females.

Genital TB can also result in infertility. In males, there is evidence for reduced semen volume, chemical alterations to the seminal fluid and a decline in sperm count and mobility; obstruction of the genital ducts and subsequent sterility are common sequelae. In women, tubal closure can occur, or if not, the fallopian tubes may be so involuted that sperm die before reaching the ovum. The incidence of sterility among women with proven genital tuberculosis varies between 55% and 85%. Because of endometrial lesions, implantation is unlikely to be successful and abortion will occur; in only 2% of women treated for genital TB did a successful pregnancy result. If a pregnancy occurs it may well be ectopic.

Leishmaniasis

Much of the interest in fertility focuses on tropical and developing countries where parasitic diseases are common place. Leishmaniasis refers to a group of clinical diseases produced by protozoan organisms of the genus *Leishmania*. There are three distinct clinical forms of which kala-azar (visceral or systematic leishmaniasis) is the most important. This disease which is widely distributed throughout the world has its greatest prevalence in India and China is transmitted by the bite of sandflies (*Phlebotomus*). If kala-azar is acquired during pregnancy, then there is increased likelihood of fetal loss (Lee, 1982).

Trypanosomiasis

Two forms of human trypanosomiasis exist. The South and Central American form is called Chagas's disease and it is caused by *Trypanosoma cruzi*. In Africa, the disease is known as African sleeping sickness and the agents are *Trypanosoma brucei gambiense* and *Trypanosoma brucei rhodesiense*.

At least 25 million people are infected with *T. cruzi* with a further 65 million at risk. *T. cruzi* is transmitted by triatomid bugs which live in the crevices in the primitive dwellings commonplace in urban shanty towns. The bugs feed by sucking blood. The bites usually occur around the mouth and nose; the trypanosome penetrates through the bite wound and enters a cell where it develops into an amastigote form. The organism multiplies and forms a pseudocyst that ruptures with the release of protozoan organisms which enter the bloodstream or invade adjacent cells.

The acute form of the disease is generally found in children and only about 1% of inhabitants of endemic areas develop clinically apparent acute disease. The mortality rate in the acute phase is 10–20% with most deaths occurring because of cardiac disease or encephalitis. The idea that *T. cruzi* can cross the placenta and infect the developing fetus was first proposed by Carlos Chagas in 1911; experimental evidence was obtained in 1921 and congenital infections in humans first found in 1949. In South America between 1 and 10% of spontaneous abortions are attributed to Chagas's disease and stillbirth is also frequently found. Congenitally infected neonates tend to be preterm and small for gestational age (in one study birth weights were less than 2000 g in 80% of cases) and a high proportion die before 4 months of age (Bittencourt, 1976). In addition *T. cruzi* may be found in breast milk (Lee, 1982).

In African sleeping sickness, the infection is transmitted to humans by the bite of the Tsetse fly. The trypanosomes do not invade the cells but remain extracellular and accumulate in connective tissue. If untreated the disease is invariably fatal since invasion of the central nervous system occurs. In earlier stages the disease is associated with fevers, resulting from the successive parasitemias, and temperatures often reach 39.4–40 °C (103–104 °F) and occasionally 41.4 °C. Fevers at these temperatures may affect male fecundity since the testicular temperature may be elevated to an extent where azoospermia occurs, followed by 1–23 months of lowered sperm count (oligospermia). In women, the fevers are thought to cause spontaneous abortion, preterm labour and delivery or stillbirth. Prematurity is also likely in severe anaemia and this is a common phenomenon in African sleeping sickness.

Malaria

Malaria is one of the most important of the protozoan infections with over 200 million people suffering malarial attacks each year. The disease is caused by four species of a protozoan parasite, *Plasmodium*, of which *P. falciparum* is the most life threatening.

There is little or no evidence that malaria has an impact on coital ability although the fevers and anaemia associated with the disease are likely to lower the likelihood of coitus. In males the high fevers are likely to lead to azoospermia with a further period of up to 2 months of lowered sperm count (oligospermia). Azoospermia occurs if the scrotal temperature exceeds 37°C for more than 45 minutes. Since men living in hyperendemic areas suffer, on average, from two malarial attacks each year (Eaton & Mucha, 1971), fertility is depressed for between 2 and 4 months each year. Gray (1977) suggested that women may be unable to conceive during a malarial attack, but there is no direct evidence to support this notion.

It has been found by both Bruce-Chwatt (1957) and Gilles, Lawson and Sibelas (1969) that (a) malaria is more frequent in pregnant than in non-pregnant women (b) the severity of attacks with *P. falciparum* are more severe in the pregnant than in the non-pregnant patient and (c) primiparous women have increased susceptibility. The high fevers associated with malaria (and typhoid fever, influenza and pneumonia) can cause spontaneous abortion and fetal mortality rates, predominantly in the first trimester, range from 14–60%.

Anaemia is commonly associated with *P. falciparum* malaria, and severe maternal anaemia may result in second trimester abortion or stillbirth. In addition, the placenta may be infected with malarial parasites leading to intrauterine growth retardation with the likelihood of elevated perinatal mortality rates. Reinhardt (1978) found higher antibody titres against *Plasmodium* in mothers with low birth weight babies (< 2500 g), pre-term deliveries (< 37 weeks) and small-for-gestational age infants than did control patients. In addition, there is a risk of congenital malaria and Reinhardt has suggested three mechanisms by which this could occur (a) transplacental passage of parasites (b) inoculation of parasites from maternal to fetal blood via skin abrasions at the time of delivery and (c) passage of parasites into amniotic fluid.

Helminths

Helminths or worms are prevalent particularly in tropical regions. It is estimated that *Ascaris* (roundworm) and *Enterobius* (pinworm) account for 1 billion infections each, *Trichuris* (whipworm) and hookworms 500 million each and schistosomes and filariae 250 million each.

Adult *Ascaris* have been reported as invading the female genital tract (Sterling & Gray, 1936) and causing tuboovarian abscess. Chu *et al.* (1972) reported a case of congenital Ascaris infection with an infant delivered in association with 12 adult *Ascaris* worms. Migrant pinworms have been found to ascend the vagina producing vaginitis and on one occasion pelvic inflammatory disease resulted (Brooks, Goetz & Plauche, 1962). Hookworm can lead to iron deficiency anaemia which may have implications for maternal health and wellbeing. Trichuris is not a significant threat to pregnant women or their fetuses.

There are two major groups of filarial infections, bancroftian filariasis (caused by *Wuchereria bancrofti*) and malayan filaraisis (caused by *Brugia malayi*). They present similar clinical infections and are transmitted to humans by mosquitoes. Following a bite from an infected mosquito the larvae pass into the lymphatics and lymph nodes where they mature into adult worms. Female worms discharge microfilariae into the bloodstream. These microfilariae can invade the placenta and fetus especially in deliveries occurring during the night. Carayon, Brenot and Camain (1967) have reported the presence of chronic bancroftian filariasis in the fallopian tubes or ovaries, with concomitant infertility.

The male genitals are frequently the site of inflammatory lesions of filariasis with inflammation of the spermatic cord, epididymis and testis. These inflammations are likely to make coitus very painful. In the chronic stage, elephantiasis can lead to greatly enlarged penis or scrotum which would make coitus very difficult if not impossible. Elephantiasic scrotum are frequently associated with degenerative changes in the testis and the absence of spermatogenesis has been reported (Spingarn & Edelman, 1965). In females, vulval elephantiasis may interfere with coitus and labour may be obstructed.

Schistosomiasis is one of the most widespread parasitic diseases which has a complex lifecycle involving a snail intermediate host. The adult flukes are found either in the mesenteric and hemorrhoidal veins (*S. mansoni*), mesenteric and portal, veins (*S. japonicum*) or bladder venous plexus (*S. haematobium*). Clinical symptoms of the disease result from the inability of the body to excrete all the eggs released by the fluke. The host response is to develop a granuloma around each egg. With resolution of this lesion, collagen deposition and fibrosis occur thus producing clinical signs.

The effect of schistosomiasis on male fecundity has been debated for quite sometime. The early view was that azoospermia could result from infestation of the urogenital tract by *S. haematobium* (Aal *et al.*, 1975), but this is now no longer thought to be correct. In females, the genital tract can be infected with eggs of both *S. mansoni* and *S. haematobium*. Rosen and Kim (1974) suggested that acute and chronic inflammation of the fallopian tubes can result in ectopic pregnancy and infertility. Inflammation of the cervix, vagina and vulva may also occur, making coitus painful or making vaginal delivery difficult.

The impact of sexually transmitted diseases on subfecundity

There are over 20 pathogens which are sexually transmitted. These pathogens include bacteria, viruses, fungi, protozoa and ectoparasites. There is currently an epidemic of sexually transmitted diseases (STDs) and to the classic five diseases of gonorrhoea, syphilis, chancroid, lympho-granuloma venereum and granuloma inguinale have to be added diseases such as nongonoccocal urethritis, epididymitis, vaginitis, cervicitis, pelvic inflammatory disease, hepatitis, genital oncogensis and severe immunosup-pression such as AIDS.

There has also been a change of emphasis in the field of STDs (Cates, 1987) with less concern with traditional venereal disease such as gonor-rhoea and more emphasis on syndromes associated with *Chlamydia trachomatis*, herpes simplex virus (HSV), human papilloma virus (HPV) and recently to concern with the fatal disease AIDS, caused by human immunodeficiency virus (HIV). This change of emphasis is not associated with the emergence of new pathogens (except for HIV) but with increased knowledge and awareness resulting from improvements in laboratory diagnostic techniques. Thus it is now known that increasing numbers of young sexually active adults are at risk for STDs with increases in age adjusted incidence of newer STDs such as *C. trachomatis*, HSV and HPV. Furthermore it is also recognized that the newer STDs are associated with incurable and fatal conditions – HIV and AIDS, HPV associated with genital cancers, and chronic recurrent HSV.

With the exception of AIDS, STDs have more serious long-term consequences for women than men. For instance, women show an increased risk of genital cancer with HPV, loss of reproductive capability resulting from damage to the fallopian tubes with *N. gonorrhoea* and *C. trachomatis*, premature and ectopic pregnancy and risk of transmission of serious or fatal infections to the fetus or newborn (syphilis, HSV, C. trachomatis and hepatitis B virus).

Gonorrhoea

The agent for gonorrhoea is the bacterium *Neisseria gonorrhoeae*. It is the most commonly reported communicable disease in the USA and in 1992 approximately 700 000 cases of gonorrhoea were reported but this figure rises to between 1.5 and 2 million cases when the severe under-reporting is taken into account. Overall, more males are affected than females (1.5:1) but in young teenagers the ratio is reversed.

A number of factors have been shown to associate with the risk of becoming infected. These include age, sex, ethnicity and number of sexual partners; use of the contraceptive pill may enhance transmission since the pill hormones increase the pH of the vagina thereby creating an alkaline environment particularly favourable to the gonococcus.

The transmission of gonorrhoea is almost entirely as a result of sexual contact (although in the wet tropics the bacteria can survive for 3–4 hours outside the body, and there is the possibility of transmission to children sharing bed linen or clothing with an infected adult). The female is at greater risk of infection than the male; the risk of a male becoming infected after one sexual act with an infected female is between 20 and 25% while the risk of transmission from male to female is put at 80–90%.

Gonorrhoea causes specific inflammation of the mucous membranes of the genitourinary tract. In males it leads to dysuria (painful urination) and increased frequency of urination. If untreated, complications can occur: epididymis and urethral stricture leading to azoospermia. In women, the majority remain asymptomatic; others show vaginal discharge, dysuria, abnormal bleeding and pelvic discomfort. Complications arise because gonorrhoea does not remain confined to urethra and cervix but spreads upward to involve the uterus, fallopian tubes and perhaps peritoneal cavity.

Supracervical infection is called acute pelvic inflammatory disease (PID) or acute salpingitis (PID refers to the acute clinical syndrome attributed to the ascending spread of microorganisms from the vagina and endocervix to the endometrium. If associated with gonococcus it is referred to as gonoccocal PID, otherwise nongonococcal PID). PID can lead to tubal inflammation, narrowing of the tubal lumen and the build up of fibrous tissue resulting in tubal occlusion. It is estimated that one episode of PID leads to tubal occlusion in 12.8% of cases; 2 episodes to 35.5%, 3 or more episodes to 75% occlusion. The likelihood of developing PID depends on the type of contraceptive used; PID is higher in IUD users and in women never pregnant and in those who have just experienced abortion or childbirth.

In addition, there are important affects of gonococcal infection in pregnancy and the neonate. Amniotic infection syndrome due to gonococcal infection leads to placental, fetal membrane and umbilical cord inflammation resulting in an increased incidence of premature rupture of the membranes, preterm delivery, maternal postpartum sepsis. In addition, higher incidence of intrauterine growth retardation has been observed in gravid women with gonococal infection.

Syphilis

The name syphilis comes from a poem by the Italian pathologist Hieronymous Fracastorius in which the mythical shepherd Syphilis is afflicted with the disease as punishment for cursing the gods. In the USA there are about 120 000 cases/annum. In developing countries accurate rates are not available.

Syphilis is caused by a spirochete, *Treponema pallidum*. Acquisition is generally through sexual contact, except for congenital syphilis. A chancre appears at a primary lesion. If untreated, resolution occurs within 3–6 weeks and a secondary stage follows lasting between 2–6 weeks. This stage is also self-limiting and is followed by a latent phase with no clinical manifestations. Individuals with primary, secondary or early latent (up to 1 year) syphilis are capable of transmitting syphilis to susceptible hosts. Without therapy about one-third of patients develop tertiary syphilis with progressive damage to CNS, cardiovascular system and musculoskeletal system.

There is no evidence of any link between syphilis and coital inability. Those with neurosyphilis frequently become impotent, but as this tends to be at the end of the reproductive period there is little or no impact on population fecundity. Syphilis is also not an important cause of conceptive failure. Instead, the main effect of syphilis is in increasing the risk of pregnancy loss (although the outcome is highly variable) with termination as a second or third trimester abortion, stillbirth or live birth.

Congenital syphilis can also occur and the congenitally infected infant may show signs of infection immediately after birth and die neonatally or may appear normal at birth but develop signs and symptoms later. It has been shown that *T. pallidum* can be transferred across the placenta and infect fetus as early as 8 weeks gestation (Harter & Benirschke, 1976). Women with primary or secondary infection are more likely to transfer infection to their offspring than those with latent infection.

Chlamydia trachomatis

Chlamydia trachomatis has emerged as one of the most common, perhaps most common, sexually transmitted organisms. In the USA an estimated 4 million new cases occur each year. Between 4 and 5% of sexually active women carry chlamydia in their cervix and of the 1 million women who acquire PID each year between 20% and 50% are associated with *C. trachomatis*. High risk groups are readily identifiable; these include age, socioeconomic status and the number of sexual partners. Lower socioeconomic status has been associated with an increased risk for chlamydial infection and chlamydial infection rates are inversely related to age and directly related to number of sexual partners. Young women using oral contraceptives are at greater risk for cervical chlamydial infection than are women using other types of contraception.

Chlamydia is the major aetiologic agent for non-gonoccocal urethritis and postgonococcal urethritis and can give rise to epididymitis in young men. In addition, in untreated women it can spread to the upper genital tract with subsequent infertility and ectopic pregnancy. A child born to a woman with a chlamydial infection of the cervix has a 60–70% risk of acquiring the infection via vertical transmission during the passage through the birth canal. Somewhere between 25% and 50% of exposed children will develop chlamydial conjunctivitis in the first two weeks of life and 10–20% will develop chlamydial pneumonia within 3–4 months after birth.

Herpes simplex virus (HSV)

There are two forms of this virus, HSV-1 causes blisters on the mouth while HSV-2 is the genital form which is spread venearally. Genital herpes can be recurrent disease. HSV-2 might also be oncogenic but it is now thought that human papilloma virus (HPV) is the more likely in cervical carcinoma. If the sores from genital herpes are on the penis or the vulva, then intercourse can become impossible; coital ability returns when the lesions have healed. Maternal infection has an adverse effect in early pregnancy with an estimated threefold increase in the likelihood of spontaneous abortion.

AIDS

So much has been written about AIDS, only a short summary will be given here. AIDS, or acquired immunodeficiency syndrome first appeared in humans in 1976 but was not recognized as a separate disease until 1981.

Table 7.5. *Diseases associated with subfecundity*

Summary table		
Male		
Coital inability	Conceptive failure	
Genital TB	Genital TB	
	African trypanosomiasis	
Female		
Coital inability	Conceptive failure	Pregnancy loss
Genital TB	Genital TB	Genital TB
		Kala-azar
		Chagas's
	African trypanosomiasis	African trypanosomiasis

Since then it has reached epidemic proportions with about 74 000 new cases in the USA in 1991 of which 7200 were women and 1000 children. The disease agent is human immunodeficiency virus (HIV) but it was not so called until 1986.

Heterosexual acquired AIDS is increasing rapidly up from 4% in 1989 to over 10% in 1993. Sex with an HIV infected male accounted for 36% of all AIDS cases in US adult and adolescent women in 1981, and for 50% of cases reported in 1992. Transmission from male to female and female to male are both possible and there is no obvious difference in rates. The proportion of partners infected is lower for partners of haemophiliacs or transfusion recipients. The risk of transmission varies greatly in studies involving partners of IVDUs, bisexual men, or mixed group with a range of 0–86%.

Maternal child transmission of HIV occurs and the rate is put at somewhere between 20 and 50%. Pregnancy may accelerate the progression of HIV infection due to reduced T_4 and lymphocytic responses. Initial studies also showed HIV infection was associated with an increased risk for adverse pregnancy outcome but later studies have failed to confirm this finding.

Conclusions

This chapter has shown that the relationship between disease and subfecundity can be considered under three broad headings of coital inability, conceptive failure and pregnancy loss (see Table 7.5 for a summary). Although absolute coital inability is rare, a number of diseases lead to obstruction (partial or complete) of the vagina, genital deformities (e.g. penile elephantiasis) or result in impotence. The more usual sequela is difficult or painful coitus (dyspareunia) which, although usually only of a temporary nature, is likely to lead to reduced coital activity.

Conceptive failure can occur in both males and females. In males, high fevers from malaria and filariasis can lead to azoospermia followed by a further period of up to 2 months of oligospermia. The recurrent bouts of fever in hyperendemic malarial areas (2 bouts a year) and the more frequent filarial fevers attacks (up to six per year) are likely to be an important cause of male infertility in Africa (McFalls & McFalls, 1984). In addition, genital infection or inflammation can lead to fibrotic occlusion of the epididymis (e.g. tuberculosis and gonoccocal epididymitis). In females conceptive failure has a variety of causes: immunological, endocrinological or damage to the ovaries or fallopian tubes. An example of an immunological problem is of women with genital schistosomiasis who are thought to produce antispermal antibodies, thereby lowering conceptive ability. It is well known that hypothalamic damage occurs in women in the advanced stages of African sleeping sickness. Such women suffer from endocrine imbalance because of the hypothalamic–pituitary–gonadal axis impairment. Finally, damage to ovaries and/or tubes can occur and is a common sequela of PID, use of an intrauterine device (IUD) or following schistosomiasis or filariasis.

Pregnancy loss may occur at any time between conception and birth through a variety of causes (McFalls & McFalls, 1984). Primary causes of pregnancy loss include damage or obstruction to the fallopian tubes, uterus, vagina or pelvic bones. Secondary causes result from endocrine problems, anaemia, transplacental transmission of infectious or placental parasitization.

Although it is customary to differentiate between developing and developed countries, there is no doubt that sexually transmitted diseases are important contributors to subfecundity in all countries. Concerns with the traditional venereal diseases such as gonorrhoea and syphilis have been replaced by emphasis on syndromes associated with *C. trachomatis*, herpes simplex virus, human papilloma virus and with the fatal disease AIDS, caused by human immunodeficiency virus (HIV). Even so, one cannot ignore the contribution to population subfecundity by infectious diseases whose foci are in the developing countries, such as schistosomiasis, malaria, filariasis, African sleeping sickness and Chagas' disease.

REFERENCES

Aal, H., El Atribi, A., Abdel Hafiz, A. & Aidaros, M. (1975). Azoospermia in bilharziasis and the presence of sperm antibodies. *Journal of Reproduction and Fertility*, **42**, 403–12.

Abioye, A. A. (1973). Fatal ameobic colitis in pregnancy and the puerperium: a new clinico-pathological entity. *Journal of Tropical Medicine and Hygiene*, **76**, 97–100.

Bittencourt, A. L. (1976). Congenital Chagas's disease. *American Journal of Disease of Childhood*, **130**, 97–103.

Bongaarts, J. & Potter, R. G. (1983). *Fertility, Biology and Behavior*. New York: Academic Press.

Brooks, T. J., Goetz, C. C. & Plauche, W. C. (1962). Pelvic granuloma due to *Enterobius vermicularis*. *Journal of the American Medical Association*, **179**, 492.

Bruce-Chwatt, L. J. (1957). Malaria in African infants and children in Southern Nigeria. *Annals of Tropical Medical Parasitology*, **19**, 173–200.

Carayon, A., Brenot, G. & Camain, R. (1967). Vingt lesions tubo-ovariennes de la bilharziose et de la filariose de Bancroft. *Bulletin de La Social Medicine Afrique Noire Language Français*, **12**, 464–73.

Cates, W. Jr (1987). Epidemiology and control of sexually transmitted diseases: strategic evaluation. *Infectious Disease Clinical in North America*, **1**, 1–23.

Chu, W., Chen, P., Huang, C. & Hsu, C. (1972). Neonatal Ascaris. *Journal of Pediatrics*, **81**, 783–5.

Czeizel, E., Hancsok, M. & Palkowich, I. (1966). Possible relation between fetal death and *E. histolytica* infection of the mother. *American Journal of Obstetrics and Gynecology*, **96**, 264.

Davis, K. & Blake, J. (1956). Social structure and fertility: an analytic framework. *Economic Development and Cultural Change*, **4**, 211–32.

Eaton, J. & Mayer, A. (1954). *Man's Capacity to Reproduce. The Demography of a Unique Population*. Glencoe, IL: Free Press.

Eaton, J. & Mucha, J. (1971). Increased fertility in males with sickle cell trait. *Nature*, **231**, 456.

Fivnat (1993). French national IVF registry: analysis of 1986 to 1990 data. *Fertility and Sterility*, **59**, 587–95.

Frank, E., Anderson, C. & Rubinstein, D. (1978). Frequency of sexual dysfunction in normal couples. *New England Journal of Medicine*, **299**, 111–19.

Gilles, H. M., Lawson, J. B. & Sibelas, M. (1969). Malaria, anemia and pregnancy. *Annals of Tropical Medical Parasitology*, **63**, 245–63.

Guay, R. (1977). Biological factors other than nutrition and lactation which may influence natural fertility: a review. In *Natural Fertility* ed. H. Leridon and J. Menken, Ordina, Liège, Belgium.

Guinness Book of Records 1993. (1992). Guinness Publishing Limited.

Harter, C. A. & Benirschke, K. (1976). Fetal syphilis in the first trimester. *American Journal of Obstetrics and Gynecology*, **124**, 705–11.

Kreutner, A. K., Del Bene, V. E. & Amstey, M. S. (1981). Giardiasis in pregnancy. *American Journal of Obstetrics and Gynecology*, **140**, 895–9.

Lee, R. V. (1982). Parasitic infections. In *Medical Complications during Pregnancy* ed. G. N. Burrows and T. F. Ferris, Philadelphia: W. B. Saunders.

McFalls, J. A. & McFalls, M. H. (1984). *Disease and Fertility*. New York: Academic Press.

Medical Research International (1990). The American Fertility Society special interest group: *in vitro* fertilization/embryo transfer in the United States: 1988 results from the National IVF–ET Registry. *Fertility and Sterility*, **51**, 13.

Masters, W. & Johnson, V. (1970). *Human Sexual Inadequacy*. Boston: Little, Brown.

Mosher, W. (1982). Infertility trends among US couples: 1965–1976. *Family Planning Perspectives*, **14**, 22–35.

Mosher, W. & Pratt, W. (1982). Reproductive impairments among married couples: United States. Vital and Health Statistics, National Center for Health Statistics, Public Health Service. Washington DC. USA: US Government Printing Office.

Reinhardt, M. C., Ambroise-Thomas, P., Cavallo-Serra, R., Meylan, C. & Gautier, R. (1978). Malaria at delivery in Abidjan. *Helvetica Paediatrica Acta*, **33**, Supplement 41, 65.

Reinhardt, M. C. (1980). Effects of parasitic infection in pregnant women. In *Perinatal Infections*. Ciba Foundation Symposium 77. Amsterdam: Excerpta Medica.

Romaniuk, A. (1968). The demography of the Democratic Republic of the Congo. In *The Demography of Tropical Africa* ed. W. Brass, pp. 241–341. Princeton: Princeton University Press.

Rosen, Y. & Kim, B. (1974). Tubal gestation associated with *Schistosoma mansoni* salpingitis. *Obstetrics and Gynecology*, **43**, 413–21.

Spingarn, C. & Edelman, M. (1965). Parasitic diseases in relation to pregnancy. In *Medical, Surgical, and Gynecologic Complications of Pregnancy* ed. J. Rovinsky and A. Guttmacher, Baltimore: Williams and Wilkins.

Sterling, R. & Guay, A. J. L. (1936). Invasion of the female generative tract by Ascaris lumbricoides. *Journal of the American Medical Association*, **107**, 2046–7.

Part III:
Metabolic and energetic aspects of
regulation

8 Metabolic adaptation in humans: does it occur?

P. S. SHETTY

Introduction

An adaptive response is an inevitable consequence of sustained perturbation in the environment and may be genetic, physiological and or behavioural. They are not completely separate entities as they interact with each other at several levels and are not without cost to the organism. Every adaptation has its cost and there is no such thing as a 'costless' adaptation. The processes and the costs involved may be: overt or covert, reversible or irreversible and transient or permanent. An adaptive response is a homeostatic response. However, the term 'homeostasis' pertains to maintenance of the constancy of the *internal milieu* when changes occur within the acceptable range of physiological variability. A sustained perturbation in any one direction will result in responses which will further the survival of the individual and help maintain homeostasis but at a cost to the organism. A homeostatic response may neither have additional costs to the organisms nor lead to compromise in its function, capability, or performance, while an adaptive response may do both in order to further the survival of the individual. Adaptation is a relatively slow process and should be distinguished from the rapid regulatory role of homeostatic mechanisms.

Adaptations in energy metabolism of humans

It has been suggested that the energy metabolism of individuals is more variable and adaptable than previously believed, and that allowances need to be made for this when arriving at estimates of human energy requirements. Several important publications have drawn attention to the possibility of such physiological variability in energy utilization between individuals (Durnin *et al.*, 1973; Edmundson, 1980) and within individuals (Sukhatme & Margen, 1982). Norgan (1983) has critically evaluated the

125

fourfold evidences that have been adduced for this variation which is purported to result in adaptation in human energy metabolism. These include: (i) in any group of 20 or more similar individuals, energy intake can vary as much as twofold (Widdowson, 1962); (ii) large numbers of apparently healthy active adults exist on lower than required energy intakes (Durnin, 1979); (iii) the efficiency of work and work output is variable per unit energy intake (Edmundson, 1979) and (iv) the observations based on studies of experimental or therapeutic semi-starvation (Keys *et al.*, 1950; Grande, Anderson & Keys, 1958; Apfelbaum, 1978) and overfeeding of humans (Sims, 1976; Norgan & Durnin, 1980). Differences in body size, levels of physical activity and systematic errors in the estimation of energy intakes may provide explanations for most of these observations (Norgan, 1983). However, what is implied or explicitly stated by the proponents of metabolic adaptation is that metabolic efficiency and mechanical work efficiency of the individual are variable and show an adaptation to variations in the levels of energy intake. The decrease in oxygen utilization of the residual active tissue mass of an individual during experimental (or therapeutic) semi-starvation (Keys *et al.*, 1950) constitutes the most important biological argument for metabolic adaptation manifested by an enhanced metabolic efficiency in chronic undernutrition resulting from lowered energy intakes.

Since, the focus of this chapter is to examine the evidence for metabolic adaptation in humans, only those areas that are relevant to developing this argument will be discussed. The two major areas that will be examined are (i) the metabolic responses that occur during acute energy restriction in humans and the physiological mechanisms that are involved in this response and (ii) the metabolic changes that are seen in individuals who are chronically undernourished as a result of long standing low energy intakes. Since the Symposium focuses on reproduction and fertility, the question whether metabolic adaptation occurs during pregnancy and lactation will also be considered.

Metabolic adaptation during acute energy restriction in previously well nourished adults

Reduction in metabolic rates is a constant finding during energy restriction in man; either experimentally induced or during therapeutic energy restriction. Apfelbaum (1978) summarized the studies in human subjects since the beginning of the twentieth century which have demonstrated a consistent decrease in metabolic rate consequent to either spontaneous, experimental or therapeutic energy restriction. A more comprehensive review by Sims (1986) examining the responses of obese subjects to a reduction in caloric intake demonstrated most studies as showing a significant reduction in metabolic rate. This consistent finding of a reduced basal metabolic rate (BMR) during experimental semi-starvation and following therapeutic energy restriction was explained both on the basis of a decrease *per se* in the activity of the metabolically active tissues of the body and as a consequence of the loss of active tissue mass due to loss of body weight. The former response was considered as indicating an increase in the 'metabolic efficiency' of the residual active tissue mass and hence evidence of 'metabolic adaptation' to the change in the energy balance status of the individual. Taylor & Keys (1950) considered the decrease in the mass of metabolically active tissue as the main factor responsible for the reduced BMR. From an analysis of the Minnesota semi-starvation data, they concluded that about 65% of the reduction in BMR can be attributed to the shrinkage of the metabolizing mass of body cells and only about 35% to metabolic adaptation i.e. an actual decrease in the cellular metabolic rate. Grande *et al.* (1958) applied identical methods of computation to data from another series of experimental human semi-starvation studies, and they found quite the reverse distribution with the actual decrease in metabolic activity of the cells contributing to a greater percentage (i.e. 65 to 73%) of the reduction in BMR. The differences in the two experimental semi-starvation studies were then explained by the obvious differences in the duration of energy restriction in the two studies. In comparison to the long-term Minnesota semi-starvation study of 24 weeks, the experiments of Grande and his colleagues (1958) did not exceed energy restriction beyond 3 weeks. Nevertheless, the decrease in metabolic rate expressed per unit weight of active tissue, (an expression considered to be *sine qua non* with metabolic adaptation) in the short term energy restriction study by Grande *et al.* (1958) was of the same order of magnitude as that observed in the Minnesota studies after 24 weeks of energy restriction (Keys *et al.*, 1950). Recalculating the data from the two separate experimental semi-starvation studies by Keys *et al.* (1950) and Grande *et al.* (1958); James and Shetty (1982) were able to show that the early fall in BMR seen during energy restriction is mainly accounted for by enhanced metabolic efficiency (Table 8.1). This reduction in BMR per unit active tissue seen in the first 2 weeks of energy restriction by Grande *et al.* (1958) remained essentially unchanged over the subsequent 22 weeks of semi-starvation (Keys *et al.*, 1950). However, the greater contribution to the fall in BMR in the long-term adaptation to prolonged energy restriction was the slow decrease in the total mass of active tissues as a result of body weight loss. Comparison of

Table 8.1. *Changes in body weight, active tissue mass (ATM) and basal metabolic rate (BMR) following short-term and long-term semi-starvation in humans*

	Semi-starvation			
	Short term		Long term	
	Baseline	Day 14	Baseline	Day 168
Body weight (kg)	71.6	65.4	67.5	51.7
ATM (kg)	44.9	42.2	38.8	28.7
BMR (MJ/d)	7.3	5.7	6.6	4.2
(kCal/d)	1742	1370	1575	1004
Decrease in BMR				
kJ/d		21.4%		36.3%
kJ/kg ATM		16.3%		13.8%

Data from experiments conducted on human subjects by Grande, Anderson & Keys (1958) and long term by Keys *et al.* (1950). Both sets of experiments had 12 subjects each. ATM = active tissue mass estimated from body weight minus sum of fat, bone mineral and extracellular fluid.

Adapted from James & Shetty, 1982.

more recent studies on therapeutically restricted diets in obese patients also shows the same response over the first few weeks of energy restriction although this is not an invariable finding and the degree of reduction seen varies largely with the severity of the dietary restriction and the composition of the diet.

It seems reasonable therefore to assume that the reduction in BMR during energy restriction (experimental or therapeutic) occurs in two different phases. In the initial phase (the first 2 or 3 weeks) there is a marked decrease in the BMR which is not attributable to the changes in body weight or body composition. This decrease in BMR per unit active tissue is a measure of the increase in 'metabolic efficiency' of the active tissue mass of well nourished individuals (lean or obese) who are energy restricted and an indication of the existence of metabolic adaptation. With continued energy restriction, the lowered level of cellular metabolic rate remains nearly constant and any further decrease in BMR is accounted for by the loss of active cellular tissue. Thus, the longer the duration of energy restriction, the more important becomes the contribution of decreased body tissues to the reduction in BMR. This reduction in lean body tissue with prolonged energy restriction is considered to be a passive process and a consequence of body tissues being used as substrates and metabolic fuel.

Physiological mechanisms that enhance metabolic efficiency during energy restriction

The physiological mechanisms involved in the 'active processes' resulting in an increase in metabolic efficiency and aimed at reducing cellular metabolic rate are not well understood. Several factors such as hormonal and substrate alterations may operate and interact to influence this metabolic adaptive process (Shetty, 1990). Reduction in this component is manifest in the early phase of semi-starvation and hence the related physiological mechanisms must be apparent during the first 2 to 3 weeks of energy restriction. The regulatory mechanisms that contribute to this adaptative response which decreases the rate of cellular metabolism are endocrine in nature. Several hormones are now known to be sensitive to changes in the levels of energy intake, the dietary composition and the status of energy balance of the individual. Changes in sympathetic nervous system (SNS) activity and catecholamines, alterations in thyroid hormone metabolism and changes in pancreatic peptide hormones such as insulin and glucagon play an important role in the metabolic response to energy restriction. These changes are not only aimed at lowering the metabolic activity of the active cell mass but are also essential for the orderly mobilization of endogenous substrates and fuels during a period of restricted availability of exogenous calories.

Activity of the sympathetic nervous system during energy restriction

The most important evidence linking energy intake with sympathetic nervous system (SNS) activity has been provided by a series of animal experiments carried out by Landsberg and Young (1978). These include a rapid decrease in the turnover of catecholamines in cardiac, pancreatic, liver and other tissues of 48-hour fasted rats on adequate intakes of fluid and electrolytes; while overfeeding normal and previously fasted rats by providing them free access to sucrose solutions increased the turnover of catecholamines in cardiac and other tissues. These series of studies by Landsberg and Young (1983) linked changes in catecholamine turnover with alterations in energy intake, more specifically carbohydrate intake. This concept of a reduced catecholaminergic drive during energy restriction developed by Landsberg and Young (1978) was counter to many of the accepted ideas on the control of substrate mobilization during starvation. Traditionally, the increase in lipolysis, maintenance of glucose homeostasis and the increase in glucagon output on fasting have been considered as being the result of an enhanced sympathetic drive during energy restriction. Emerging evidence shows that these responses are not dependent on an increased catecholamine activity. Adrenalectomy or adrenergic blockade does not prevent an increase in free fatty acid output on fasting while other studies summarized by Jung, Shetty and James (1980*a*) suggest that the adrenergic component of lipolysis rapidly declines with energy restriction, and that the lipolytic activity associated with energy restriction appears to be under the dominant control of declining plasma insulin levels.

Investigations in human subjects have also shown that energy restriction is associated with changes in the SNS activity. Several studies have shown that semi-starvation in humans results in a decrease in daily urinary excretion of catecholamines, a lowering in their circulating plasma levels and a decrease in the urinary excretion of metabolites of catecholamines which seem to be closely associated with the decrease in BMR per kilogram of active tissue mass that occurs during energy restriction (Shetty, 1980). Refeeding of previously energy restricted individuals produced a dramatic rise in plasma catecholamine levels and the urinary excretion of their metabolites within 72 hours. These studies confirm that the activity of the human SNS is responding to alterations in energy intake and may be responsible for many of the beneficial effects of energy restriction such as a reduction in blood pressure. As in the case of experimental animals, these responses in humans have also been shown to be specific to the carbohydrate content of the diet. There thus appears to be a similarity in the responses of the SNS activity and insulin secretion in that they are most sensitive to the carbohydrate content of the diet both during underfeeding and overfeeding. The role of SNS drive in the adaptive component of BMR is further supported by studies using *Beta* adrenergic blockade. During weight maintenance, the BMR of obese subjects reduced by 8.7% following the administration of propranolol, a *beta* adrenergic blocker (Jung *et al.*, 1980*a*) and this reduction in BMR was comparable to the reduction seen following energy restriction alone for 3 weeks (9.1%). Similar doses of propranolol produced only a <2% drop in BMR after a week of energy restriction.

Thyroid hormones and energy restriction

Thyroid hormones are important components of the metabolic adaptation associated with energy restriction and overfeeding. Changes in the thyroid hormone status seen in response to restricted intakes of energy has been well reviewed in recent years (Jung, Shetty & James, 1980*b*; Danforth & Burger, 1989) and are briefly summarized here. Energy restriction does not seem to affect the circulating levels of total T_4 (free and bound) and there is no evidence of any change in the production rates of T_4. When T_4 levels show a tendency to fall they return to prefasting levels as the fast continues. Free T_4 may be transiently elevated during starvation and this is secondary to the inhibitory effects of a high plasma free fatty acid concentration on the plasma protein binding of T_4. There is however, a rapid decline in serum T_3 and free T_3 during fasting or partial energy restriction and this response of T_3 to caloric restriction seems to be most specific to the restriction in carbohydrate intake (Danforth, 1986). The decrease in T_3 is rapid (within 24 hours) and usually reaches levels 40–50% below normal within 3 to 4 days. The drop in T_3 is not due to changes in T_3 clearance rates and is attributed to a reduction in the synthesis of T_3 by the peripheral conversion of T_4 to T_3. Associated with the reduction in T_3 there is an increase in reverse T_3 (rT_3) and free rT_3 during fasting. The increase in rT_3 is progressive, beginning much later than the fall in serum T_3 that occurs during energy restriction. After 2 to 3 weeks of fasting or semi-starvation, the rT_3 returns to normal levels when the concentrations of T_3 are being maintained at a low level. The peripheral kinetics of T_3 and rT_3 seem to be dissociated. T_3 production rates decrease during starvation while the transient changes in rT_3 depend not on changes in production but on decreased clearance rates of rT_3. This is partly due to reduced hepatic deiodinase which catabolizes rT_3 (Danforth & Burger, 1989). However, the catabolic rate of rT_3 is much higher than the rate of T_3 breakdown, thus accounting for the smaller changes in the circulating concentrations of rT_3 as compared to the plasma changes in T_3. These reciprocal changes in T_3 and rT_3 concentrations during energy restriction are bound to contribute to the reduction in metabolic rate. The tissue responsiveness to T_3 may not be altered since exogenous administration of T_3 increases the metabolic rate and reverses the BMR fall during energy restriction and weight loss. There is evidence, however, which suggests that prolonged energy restriction can decrease the nuclear T_3 receptors; this reduction in receptor numbers is not necessarily a direct consequence of the fall in T_3 levels. There is also a reduction in the binding capacity of nuclear T_3 receptors which seems to be an independent but synergistic adaptation to energy restriction.

The interactions between thyroid hormones and catecholamines may influence their mutual roles in regulation of thermogenesis. Thyroid hormones are the principal regulators of basal metabolic rates, while the regulation of adaptive thermogenesis is mediated largely by catecholamines. However, thyroid hormones do have a permissive and synergistic role in the adaptive component. Administration of T_3 even in

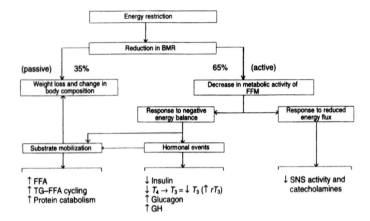

Fig. 8.1. Possible mechanisms involved in the adaptive reduction in basal metabolic rate (BMR) during short-term energy restriction. FFA = free fatty acids; FFM = fat free mass; TG = triacylglycerols; T_4 = thyroxine; T_3 = tri-iodothyronine; rT_3 = reverse T_3; GH = growth hormone; SNS = sympathetic nervous system. (Shetty, 1990.)

low doses can prevent the drop in metabolic rate seen during energy restriction (Shetty, 1980). Administration of catecholamines fail to stimulate metabolic heat production in hypothyroid animals and SNS-mediated thermogenesis in response to carbohydrate intake also requires thyroid hormones (Rothwell, Saville & Stock, 1982).

Insulin in acute energy restriction

Insulin is another pre-eminent hormone that regulates energy metabolism. Starvation or partial energy restriction brings about a significant lowering in circulating insulin levels. This is the primary hormonal signal that allows for an orderly transition from the fed to the fasted state without the development of hypoglycemia. As the fasted state extends beyond 12 to 14 hours, the decline in circulating insulin levels results in the stimulation of amino acid mobilization, gluconeogenesis, lipolysis and ketogenesis all of which are aimed at the mobilization of endogenous fuels as substrates (Sherwin & Felig, 1987). This reduction in plasma insulin level is also bound to influence the metabolic rate at rest.

In *summary*, several physiological mechanisms, chiefly hormonal, operate to decrease the metabolic activity of the active tissue mass to enhance its metabolic efficiency (Fig. 8.1). Sympathetic nervous system activity is toned down signalled by the decrease in energy flux, while the negative energy deficit lowers insulin secretion and initiates changes in peripheral thyroid metabolism; the latter are characterized by a reduction in the biologically active T_3 and an increase in the inactive reverse T_3. The reduction in the activities of these three, key thermogenic hormones acts possibly in a concerted manner to lower cellular metabolic rate. Changes in other hormones such as glucagon, growth hormone and glucocorticoids may influence these changes and at the same time in association with the insulin deficiency will promote endogenous substrate mobilization which will lead to an increase in circulating free fatty acids and ketone bodies. The elevated free fatty acid levels, alterations in substrate recycling and protein catabolism will also influence the resting energy expenditure. What contribution reductions in Na^+-K^+ pumping across the cell membrane and futile substrate cycling (now considered possible sensitive metabolic regulators) make to the reduced energy output are not known. These hormonal and metabolic changes that accompany energy restriction aid the survival of the organism during restricted availability of exogenous calories. Hence these physiological changes may be considered as metabolic adaptations which occur in a previously well nourished individual which are aimed at increasing the 'metabolic efficiency' of the residual active tissue mass at a time of energy deficit.

Metabolic adaptation and metabolic efficiency in the basal metabolism in chronically undernourished adults

Ferro-Luzzi (1985) summarized our current thinking on the ways in which an individual who is chronically undernourished may metabolically adapt and respond to the sustained and long-term energy imbalance. Metabolic adaptation was represented as a series of complex integrations of several different processes that occurred during energy deficiency (Fig. 8.2). These processes were expected to occur in phases which could be distinguished and a new level of equilibrium was then achieved at a lower level. At this stage, individuals who had gone through the adaptive processes that occur during long-term energy deficiency, were expected to exhibit more or less permanent sequelae or costs of adaptation, which included a smaller stature and body size, an altered body composition, a lower BMR, a diminished level of physical activity and the possibility of a modified or enhanced metabolic efficiency of energy handling by the residual active tissues of the body. In this section an attempt is made to critically re-examine the existence of mechanisms that may contribute to any apparent metabolic efficiency of the tissues that may contribute to meta-

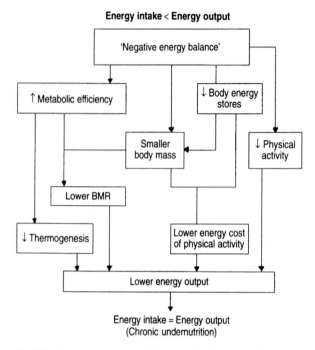

Fig. 8.2. Schematic outline of the postulated changes that lead to a new level of energy balance in chronic undernutrition. (Modified from Ferro-Luzzi, 1985.)

bolic adaptation in the chronically energy deficient or undernourished adult.

Metabolic economy in basal metabolism

It is generally assumed that metabolic adaptation occurs in the chronically undernourished state and that the physiological and metabolic responses of the adult human body with chronic undernutrition are similar to, and can be explained on the basis of, the sort of changes that occur during experimental or therapeutic semi-starvation in previously well-nourished adults. As a result, there is considerable interest to look for physiological changes suggestive of an increase in metabolic efficiency of the active tissues mass in chronically undernourished individuals in much the same way as one observes during energy restriction of well-nourished or obese subjects. Consequently, a reduction in oxygen consumption per unit of active tissue mass has been considered to be a definitive index of the existence of metabolic economy or adaptation (Waterlow, 1986). A decrease in BMR per kilogram of the active tissue mass (ATM) or fat free mass (FFM), in turn, implies that the increase in metabolic efficiency is demonstrable in the other components of energy expenditure, particularly a decrease in the energy cost of physical activity of the adapted individual as per the model developed by Ferro-Luzzi (1985) (Fig. 8.2). BMR expressed either in absolute terms or expressed per unit body surface area has always been found to be lower in chronically undernourished subjects although these same studies have failed to demonstrate any significant decrease in metabolic rate per kg ATM (Venkatachalam, Srikantia & Gopalan, 1954). Ashworth (1968), while reporting a 12% reduction in BMR in Jamaican subjects on low calorie intakes, also confirmed her inability to show evidence of enhanced metabolic efficiency and metabolic adaptation in BMR. These observations in the undernourished are in marked contrast to the invariable and constant demonstration of a decrease in BMR per kg ATM seen during experimental semistarvation in humans (Keys *et al.*, 1950; Grande *et al.*, 1958).

More recently, BMR measurements made in apparently healthy, but undernourished, labourers showed a definite reduction in the metabolic activity of the ATM which could perhaps be interpreted as indicating an enhanced metabolic efficiency, although the major share of the fall in BMR was attributable to a decrease in the total mass of lean tissues (Shetty, 1984). Recalculation of data from an earlier report by Ramanamurthy, Srikantia and Gopalan (1962) also showed that the BMR expressed per unit active tissue was considerably lower in adult, undernourished males. However, a large number of measurements made, over the last decade, in chronically undernourished subjects do not confirm the existence of an enhanced metabolic efficiency as indicated by a reduced oxygen consumption per unit of active tissue (Soares & Shetty, 1991). On the contrary, it is now observed that the BMR expressed per kg FFM is significantly higher in a large sample of chronically undernourished subjects as compared to the well nourished. This group of apparently healthy, chronically under-nourished adults had short statures, low body weights, low FFMs and low body mass indices (weight/height^2; BMI < 18). They had low energy intakes, came from poorer socio-economic groups and were urban or rural labourers. A comparable and equally large series of BMR measurements reported by Srikantia (1985) also in male individuals from lower socio-economic status showed a similar trend; BMR per unit body weight increasing as the weight for height expressed as a percentage of standard diminished. Recent BMR measurements made in rural South India by McNeill *et al.* (1987) were also comparable to those obtained both by Srikantia (1985) and by Soares and Shetty (1991). Results of these three studies with large sample sizes also provide no evidence of metabolic adaptation in individuals from poor socio-economic groups on lower planes of energy intake.

One of the serious problems in our quest to uncover the existence of metabolic adaptation in chronically undernourished adults is that in order to compare efficiencies in basal metabolism with that of the well nourished, metabolic rates have to be standardized to some estimate of the active tissue mass. Ravussin and Bogardus (1989) have provided mathematical reasons why basal metabolism should not be divided by the ATM or FFM since the relationship between BMR and ATM/FFM has a y and an x intercept significantly different from *zero* and that these intercepts must be taken into account. They believe that, because of the mathematical bias, it is incorrect to express metabolic rate per unit ATM/FFM in order to standardise values for the purpose of comparison between groups of different body weights or ATMs/FFMs. Since the use of the expression BMR per kg ATM/FFM was aimed at correcting for differences in the body sizes of groups of individuals, it is necessary to eliminate the possible methematical artefacts by resorting to Analysis of Covariance (ANACOVA) as Ravussin and Bogardus (1989) propose. Subjecting BMR data to ANACOVA, it has been shown that chronically undernourished subjects have a lower BMR when adjusted for body weight or FFM implying that there may indeed be some covert metabolic economy in the FFM even if the expression BMR per kg FFM were to fail to reveal this overtly (Soares & Shetty, 1991). However, whole body protein turnover has been estimated to contribute to as much as 35% of BMR (Jackson, 1985) and the lack of any change in protein turnover rates in these undernourished adults (Soares *et al.*, 1991) provides indirect support that the metabolic activity of the FFM is unlikely to be reduced in chronic undernutrition.

An objective assessment of all the data accumulated over the last 30 or more years hence leads to two definite conclusions:

(i) the BMRs of undernourished subjects, on an absolute basis or expressed per unit body surface area is significantly lower than that of the well nourished and this is largely accounted for by the low body weights of these individuals.

(ii) The reduced metabolic rate per unit ATM/FFM, considered hitherto as an expression of enhanced metabolic efficiency in the undernourished, possibly demonstrated in the earlier studies of the 1960s (Ramanamarthy *et al.*, 1962) and 1970s (Shetty, 1984), is not evident in the more recent BMR measurements being made over the last decade or more in individuals with more or less similar anthropometric characteristics except when adjusted for differences in body weight or FFM by Analysis of Covariance. The evidence that mechanisms of improved efficiency of energy utilization are operative in free living populations on low energy intakes but compromised nutritional anthropometry thus appears to be tenuous and contradictory. If enhanced metabolic efficiency is indeed present during chronic undernutrition, then a reduction in BMR per kg ATM/FFM is unlikely to reflect this phenomenon. Either BMR per kg ATM/FFM is not an index of metabolic efficiency or metabolic economy as is universally believed (Waterlow, 1986) or metabolic efficiency is neither a characteristic nor a constant feature of chronic undernutrition. A reduction in BMR per unit ATM/FFM is probably not *sine qua non* for metabolic efficiency of the residual tissues of the body.

Body composition and basal metabolism

One possible way in which these apparently contradictory data may be rationalized is to seek alternative explanations based on the variable changes in body composition that is likely to occur in chronic undernutrition.

Lawrence and others (1988) observed differences in BMR between Scottish, Gambian and Thai women, which could largely be explained in terms of the differences in the mass of FFM since individuals in all the three groups with similar FFMs had similar BMRs. Thus differences between the groups in BMR per kg FFM were largely explained by the between-group differences in the mass of the FFM. Within any group, the BMR per kg FFM decreased as the body weight or FFM increased. Lawrence, Thongprasert and Durnin (1988) suggested that variations in BMR per unit FFM between the heavy and light individuals in any group could indicate that the composition of the FFM was not constant. Subsequently, in a large series of autopsy studies in normal individuals it was shown that the organ mass (OM) component of FFM is related to the size of the FFM; OM:FFM ratio increasing as FFM reduces in both males and females (Garby & Lammert, 1992). Owen *et al.* (1990) and Weinsier *et al.* (1992) have respectively shown that BMR per kg body weight and BMR per kg FFM falls as body weight or FFM increases. Weinsier *et al.* (1991) have indicated that the index BMR/FFM does not take into account the fact that the metabolic rate of the FFM is not constant over a wide range of FFM and that the relative proportions of its metabolic components i.e. muscle mass vs organ mass may change as the weight of FFM changes.

In the Minnesota semi-starvation studies, at the end of 24 weeks, muscle losses were estimated at 41% whereas the reduction in ATM was only 27%

(Grande, 1964); a situation quite unlike the changes seen in acute starvation in humans. Studies examining changes in body composition of adults with naturally evolving chronic undernutrition revealed a gradation of changes related to the severity of the deficiency (Barac-Nieto *et al.*, 1978). Body cell mass (i.e. cell solids and intracellular water estimated from total body water and extracellular water) was reduced even with moderate deficiency and the muscle mass was more affected than other cells; muscle cell mass seemed to decrease linearly with the increasing severity of undernutrition, while the visceral organ and cell mass showed little change (Shetty, 1995). Body fat and body cell mass reduced by 29% while muscle cell mass decreased by 41% in severe undernutrition. Estimates of body composition indirectly estimated in chronically undernourished adults by creatinine excretion also indicate a greater reduction in muscle mass with visceral mass apparently being spared. The non-muscle mass/muscle mass ratio was 1.1 in the undernourished as compared to 0.7 in the well nourished (Soares *et al.*, 1991).

The visceral component of the FFM (liver, heart, kidney) has been estimated to utilize nearly 45% of the total oxygen consumption at rest while skeletal muscle which comprises up to 50% of the body weight contributes only 18% to the resting metabolic rate (Passmore & Draper, 1965). The combined weight of the brain and liver which accounts for 3% to 5% of the total body weight, utilizes as much as 40% of resting oxygen consumption (Keys, Taylor & Grande, 1973). Elia (1992) has estimated that 40% of the body weight of an adult man is muscle, but contributes to only 22% of BMR. Thus if the FFM had a significantly greater proportion of the metabolically active visceral mass and a reduction in the mass of the relatively less active muscle, then BMR expressed per kg FFM (or ATM) would be apparently high. This may indeed be the case with mild to moderate undernutrition in adults since muscle mass is more likely to be reduced than non-muscle or visceral mass. As undernutrition progresses, mobilisation of tissue from the visceral mass occurs in the more severe forms of energy deficiency, resulting in the BMR per kg FFM (or ATM) being reduced (Fig. 8.3). Along with these changes in body composition seen during long-term energy inadequacy, there is also a change in the extra-cellular fluid compartment which will also contribute to influence the BMR expressed per unit ATM (Barac-Neito *et al.*, 1978; Widdowson, 1985). Variations in the body composition of the chronically undernourished, more specifically the relative contributions of non-muscle and muscle mass to FFM, may account for much of the changes seen when BMR is expressed per unit ATM/FFM. If the range of body composition changes seen during the evolution of the chronically undernourished state

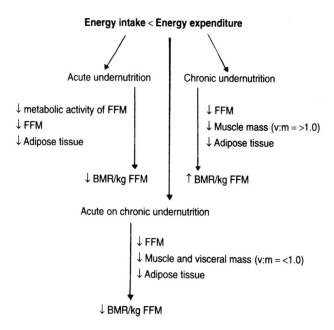

Fig. 8.3. Body composition changes seen in stages of adult undernutrition that may influence the parameter BMR per kg FFM (an indicator of metabolic efficiency). FFM = fat-free mass, V:M = visceral:muscle ratio, BMR = basal metabolic rate.

in the adult influences the parameter 'BMR per unit ATM/FFM' undoubtedly then the index BMR per unit ATM or FFM cannot truly reflect a change in the metabolic efficiency or metabolic economy of the residual active tissues of the body. It would then appear that an increase in metabolic efficiency in the BMR component of the energy expenditure, which has been hitherto considered to be the cornerstone of the beneficial, metabolic adaptation to energy inadequacy, it itself of doubtful existence.

Metabolic efficiency in substrate oxidation rates in chronic undernutrition

Respiratory Quotients(RQs) of chronically undernourished subjects shows that individuals of poor nutritional status have a higher basal, 12 hour post-absorptive fasted RQs as compared to well-nourished adults (Shetty, 1992). The higher RQs have generally been attributed to the high carbohydrate containing diets consumed by them. This seems a reasonable assumption. Their food quotients(FQs) also corroborate this with the antecedent habitual diets of these individuals being high in carbohydrate content. It seems apparent, however, that the macronutrient composition of the diet ingested as indicated by the mean FQ of the day is not truly reflected in the fasting RQ of the same day. When substrate oxidation rates are calculated during the post-absorptive, fasted state in the undernourished subjects using indirect calorimetry and urinary nitrogen excretion, it is evident that the undernourished subjects have significantly higher rates of carbohydrate oxidation and lower rates of fat oxidation in the fasted (and in the fed) state as compared to the well-nourished adult (Piers, Soares & Shetty, 1992). No differences were seen in the rates of protein oxidation; an observation that is in keeping with the evidence of similar rates of protein turnover in these subjects as compared to the well nourished (Soares *et al.*, 1991). It would appear that, in the undernourished state, the higher RQ is largely the result of selective use of carbohydrate as fuel even in the post-absorptive, fasted state. It is important to recognize that, unlike generally believed, fat is not necessarily the predominant substrate in the post-absorptive fasted state and is certainly not the preferred substrate in the chronically undernourished state. However, the selective utilization of carbohydrate illustrates how closely carbohydrate oxidation is adjusted not only to its immediate availability as demonstrated in well-nourished individuals (Flatt, 1987) but also relates probably to the antecedent habitual intakes of carbohydrate in the diets of the undernourished. The selective use of carbohydrate as fuel has obvious metabolic advantages to the undernourished individual since carbohydrate (glycogen) oxidation results in more ATP generation than iso-energetic amounts of fat or protein (Livesey, 1984; Elia & Livesey, 1992). Also the metabolizable energy equivalent, i.e. the energy equivalent of ATP gained (in kJ per mole ATP) is almost identical to that of fat (75.3 for glycogen oxidation via glycolysis and the citric acid cycle as compared to 79.2 for fat oxidation via *Beta* oxidation and the citric acid cycle (Elia & Livesy, 1992). It is hence not unlikely that the high fasting RQs of the undernourished reflect to some degree, the metabolic efficiency of the active tissues of these subjects by resorting to specific metabolic pathways that favour more efficient utilization of the available metabolic fuel. The RQ of an individual seems to reflect his body composition and more specifically the available fat stores. Flatt (1972) has suggested that the antilipolytic effect of insulin is less effective in the presence of an increased fat mass, and an increased level of insulin is thus associated with high free fatty acid levels in obesity. Since fat oxidation has been shown to be directly related to the levels of free fatty acids (Issekutz *et al.*, 1968; Groop *et al.*, 1991), it is apparent that low fasting levels of free fatty acids are likely to be associated both with the small fat mass seen in the undernourished and consequently associated with lower rates of fat oxidation. The low rates of fat oxidation will in turn contribute to the high fasting RQs. The high RQs of the undernourished may thus reflect both a high dietary intake of carbohydrate as well as a predominant dependence on carbohydrate as fuel and a reduced rate of fat oxidation in the presence of the low fat stores observed in these individuals.

It may hence be *concluded* that the BMRs of chronically undernourished adults are lower than those of well-nourished adults but not different when corrected for differences in body weight, ATM or FFM. BMR per kg ATM or FFM hitherto considered the definitive indicator of metabolic efficiency does not seem to be altered during chronic undernutrition, and there is thus no conclusive evidence of the existence of metabolic adaptation. This index does not truly reflect changes in the efficiency of energy utilization of the active tissues and is probably an artefact attributable to the changes in body composition, more specifically the disproportionate reduction in muscle tissue with a normal or even increased non-muscle or visceral organ size (possibly contributed to by the increase in number of infective episodes in the undernourished individuals), that occur during the natural evolution of the chronically undernourished state. Hence it is highly unlikely that the existence of an enhanced metabolic efficiency is demonstrable in these individuals or that metabolic adaptation occurs in the chronically under-nourished state. There is thus an urgent need to make a paradigm shift from the schema suggested by Ferro-Luzzi (1985) to explain the changes in energy metabolism that may contribute to the attaining of a new level of equilibrium during the evolution of the chronically undernourished state as a result of prolonged periods of low planes of energy intake.

Metabolic adaptation during pregnancy in humans

In the context of this symposium on 'Variability in human fertility' it seems only appropriate that the question of metabolic adaptation during human pregnancy be discussed. The hypothesis centres around the argument that the energy cost of human pregnancy can be minimized by energy-sparing metabolic adaptations when food intake is marginal or limited. Comparative studies in well-nourished women show no evidence of adaptations in BMR during pregnancy (Forsum, Sadurskis & Wager, 1988; Durnin *et al.*, 1987; van Raaij *et al.*, 1987; Piers *et al.*, 1995). This observation occurs despite evidence of very small increments in food intake during pregnancy which may not match their theoretical extra energy requirements (Durnin *et al.*, 1987). However, it has since been argued that BMR of pregnant women consistently shows an increase during pregnancy only when considered as a group and that individual women show characteristic changes in basal metabolism, which follow consistent patterns within each subject, but are highly variable between subjects (Prentice *et al.*, 1989). The percentage change in BMR above pre-pregnant BMR levels varied over a fourfold range (between 8.6 and 35.4%) in 60 Cambridge women. This between-subject heterogeneity in BMR response to pregnancy of well-nourished women was also associated with a consistent decrease in BMR per kg FFM through most of pregnancy in the grouped data. BMR/FFM changes were, however, highly variable among individuals throughout pregnancy (Prentice *et al.*, 1989). It is not unlikely that the dramatic changes in body composition that occur during the various stages of pregnancy and the technical difficulties associated with measuring FFM accurately in pregnant women may result in considerable inaccuracy in the estimate of BMR per unit active tissue that it would be well nigh impossible to use BMR/FFM as an indicator of metabolic efficiency in the pregnant state in humans.

Many women in developing countries are seemingly sustaining pregnancy on dietary intakes apparently lower than international recommendations (FAO/WHO/UNU, 1985). It has been suggested that the energy cost of human pregnancy in undernourished Gambian women is associated with metabolic adaptations that can possibly spare significant amounts of energy and may protect fetal growth (Lawrence *et al.*, 1987, Prentice & Whitehead, 1987). The hypothesis that energy-sparing adaptations in metabolism may occur during pregnancy was based on the evidence of changes in basal metabolism apparently triggered by cues from food supply (Lawrence *et al.*, 1984). A more recent, carefully conducted study in 58 Gambian women seems to show a similar degree of between subject heterogeneity in BMR changes even among Gambian women (Poppitt *et al.*, 1993). The changes in absolute BMR (i.e. in Megajoules per day) show two distinct patterns, either a consistent rise in the so-called 'energy-profligate' group or an initial decline followed by a later rise during the third trimester of pregnancy in the 'energy-sparing' group. These changes in BMR during pregnancy could not be attributed simply to changes in body weight or FFM. BMR expressed per unit active tissue showed a wide range of variation between -15.4% to $+17.1\%$. Obviously, these changes in BMR/ATM are dependent on the validity of the estimates of ATM in the field and in women in the pregnant state who manifest large changes in body composition and body water distribution. This was further complicated by the fact that there was a large inter-subject variability in the increase in body weight, lean tissue and body fat content of the Gambian women during pregnancy. It did appear that the changes in BMR in pregnancy were significantly influenced by the body weight and body composition changes with those with least increase in body weight and lean or fat tissue showing the more marked depression of BMR. The results of this study by Poppitt and others (1993) confirm that there are large inter-subject variations in BMR during pregnancy, even among women on marginal intakes of energy. Given the dual problems of marked and variable changes in body composition and the technical difficulties associated with accurate estimates of active tissue mass in undernourished, pregnant women, it would be difficult if not impossible to demonstrate the existence of metabolic efficiency using the index BMR per unit ATM/FFM, given that this index does not truly reflect at any time real changes in cellular metabolic activity. It is important, however, to note that, despite identifying these possible adaptive responses as a possible means to protect fetal growth during pregnancy, such energy-sparing adaptations when seen are definite indications of a system being stressed beyond its limits and are hence likely to influence fetal growth and pregnancy outcome such as birth weight (Poppitt *et al.*, 1994). With several repeated pregnancies, energy-sparing adaptations may impose damage not only to the fetus but also to the mother in the long term.

REFERENCES

Apfelbaum, M. (1978). Adaptation to changes in caloric intake. *Progress in Food and Nutrition Science*, **2**, 543–59.
Ashworth, A., (1968). An investigation of very low calorie intakes reported in Jamaica. *British Journal of Nutrition*, **22**, 341–55.
Barac-Nieto, M., Spurr, G. B., Lotero, H. & Maksud, M. G. (1978). Body composition in chronic undernutrition. *American Journal of Clinical Nutrition*, **31**, 23–40.
Benedict, F. G., Miles, W. R., Roth, P. & Smith, H. M. (1919). *Human Vitality and Efficiency under Prolonged Restricted Diet.* Carnegie Institute of Washington Publication No. 280, Carnegie Institute of Washington, DC.
Danforth, E. Jr. (1986). Effect of fasting and altered nutrition on thyroid hormone metabolism in man. In *Thyroid Hormone Metabolism*, ed. G. Henneman, pp. 335–58. New York: Marcel Dekker.
Danforth, E. Jr. & Burger, A. G. (1989). The impact of nutrition on thyroid hormone physiology and action. *Annual Review of Nutrition*, **9**, 201–27.
Durnin, J. V. G. A. (1979). Energy balance in man with particular reference to low energy intakes. *Bibliotheca Nutritio et Dietata*, **27**, 1–10.
Durnin, J. V. G. A., Edholm, O. G., Miller, D. S. & Waterlow, J. C. (1973). How much food does man require? *Nature*, **242**, 418.
Durnin, J. V. G. A., McKillop, F. M., Grant, S. & Fitzgerald, G. (1987). Energy

requirements of pregnancy in Scotland. *Lancet*, **ii**, 897–900.

Edmundson, W. (1979). Individual variations in basal metabolic rate and mechanical work efficiency in East Java. *Ecology of Food and Nutrition*, **8**, 189–95.

Edmundson, W. (1980). Adaptation to undernutrition: how much food does man need? *Social Science and Medicine*, **14D**: 19–126.

Elia, M. (1992). Organ and tissue contribution to metabolic rate. In *Energy Metabolism: Tissue Determinants and Cellular Corollaries*, ed. J. M. Kinney, pp. 61–79. New York: Raven Press.

Elia, M. & Livesey, G. (1992). Energy expenditure and fuel selection in biological systems: the theory and practice of calculations based on indirect calorimetry and tracer methods. *World Review of Nutrition and Dietetics*, **3**, 1–52.

FAO/WHO/UNU Expert Consultation (1985). Energy and protein requirements. *WHO Technical Report Service*, **724**, 1–206.

Ferro-Luzzi, A. (1985). Range of variation in energy expenditure and scope of regulation: In *Proceedings of XIIIth International Congress of Nutrition*, ed. T. G. Taylor and N. K. Jenkins, pp. 393–399. London: John Libbey.

Flatt, J. P. (1972). Role of the increased adipose tissue mass in the apparent insulin sensitivity to obesity. *American Journal of Clinical Nutrition*, **25**, 1189–92.

Flatt, J. P. (1987). Dietary fat, carbohydrate balance and weight maintenance: effects of exercise. *American Journal of Clinical Nutrition*, **45**, 296–306.

Forsum, E., Sadurskis, A. & Wager, J. (1988). Resting metabolic rate and body composition of healthy Swedish women during pregnancy. *American Journal of Clinical Nutrition*, **47**, 942–7.

Garby, L. & Lammert, O. (1992). An explanation for the nonlinearity of the relation between energy expenditure and fat free mass. *European Journal of Clinical Nutrition*, **46**, 235–6.

Grande, F. (1964). Man under caloric deficiency. In *Handbook of Physiology, Adaptation to the Environment*, pp. 911–37. Washington: American Physiological Society.

Grande, F., Anderson, J. T. & Keys, A. (1958). Changes of basal metabolic rate in man in semistarvation and refeeding. *Journal of Applied Physiology*, **12**, 230–8.

Groop, L. C., Bonodomma, R. C., Shank, M., Petries, A. S., De Fronzo, R. A. (1991). Role of free fatty acids and insulin in determining free fatty acid and lipid oxidation in man. *Journal of Clinical Investigation*, **87**, 83–9.

Issekutz, B., Paul, P., Miller, H. I. & Bortz, W. M. (1968). Oxidation of plasma FFA in lean and obese humans. *Metabolism*, **17**, 62–73.

Jackson, A. A. (1985). Nutritional adaptation in disease and recovery. In *Nutritional Adaptation in Man*, ed. K. Blaxter and J. C. Waterlow, pp. 111–26. London: John Libbey.

James, W. P. T. & Shetty, P. S. (1982). Metabolic adaptation and energy requirements in developing countries. *Human Nutrition: Clinical Nutrition*, **36**, 331–6.

Jung, R. T., Shetty, P. S. & James, W. P. T. (1980a). The effect of *beta*-adrenergic blockade on resting metabolic rate and peripheral thyroid metabolism in obesity. *European Journal of Clinical Investigation*, **10**, 179–82.

Jung, R. T., Shetty, P. S. & James, W. P. T. (1980b). Nutritional effects on thyroid and catecholamine metabolism. *Clinical Science*, **58**, 183–91.

Keys, A., Brozeck, J., Henschel, A., Mickelson, O. & Taylor, H. L. (1950). In *The Biology of Human Starvation*, Minneapolis: University of Minneapolis Press.

Keys, A., Taylor, H. L. & Grande, F. (1973). Basal metabolism and age of adult man. *Metabolism*, **22**, 579–87.

Landsberg, L. & Young, J. B. (1978). Fasting, feeding and regulation of the sympathetic nervous system. *New England Journal of Medicine*, **298**, 1295–301.

Landsberg, L. & Young, J. B. (1983). The role of the sympathetic nervous system and catecholamines in the regulation of energy metabolism. *American Journal of Clinical Nutrition*, **38**, 1018–24.

Lawrence, M., Lawrence, F., Lamb, W. H. & Whitehead, R. G. (1984). Maintenance energy cost of pregnancy in rural Gambian women and influence of dietary status. *Lancet*, **ii**: 363–5.

Lawrence, M., Coward, W. A., Lawrence, F., Cole, T. J. & Whitehead, R. G. (1987). Fat gain during pregnancy in rural African women: the effect of season and dietary status. *American Journal of Clinical Nutrition*, **45**, 1442–50.

Lawrence, M., Thongprasert, K. & Durnin, J. V. G. A. (1988). Between-group differences in basal metabolic rates: an analysis of data collected in Scotland, the Gambia and Thailand. *European Journal of Clinical Nutrition*, **42**, 877–91.

Livesey, G. (1984). The energy equivalents of ATP and the energy value of food proteins and fats. *British Journal of Nutrition*, **51**, 15–28.

McNeill, G., Rivers, J. P. W., Payne, P. R., deBritto, J. J. & Abel, R. (1987). Basal metabolic rate of Indian men: no evidence of metabolic adaptation to a low plane of nutrition. *Human Nutrition: Clinical Nutrition*, **41C**, 473–84.

Norgan, N. G. (1983). Adaptation of energy metabolism to level of energy intake. In *Energy Expenditure Under Field Conditions*. ed. J. Parizkova, pp. 56–64. Prague: Charles University.

Norgan, N. G. & Durnin, J. V. G. A. (1980). The effect of weeks of overfeeding on the body weight, body composition, and energy metabolism of young men. *American Journal of Clinical Nutrition*, **33**, 978–88.

Owen, O. E., Smalley, K. J., D'Alessio, D. A., Mozzoli, M. A., Knerr, A. N., Kendrick, Z. V., Kavle, E. C., Donohoe, M., Tappy, L. & Boden, G. (1990). Resting metabolic rate and body composition of achondroplastic dwarfs. *Medicine*, **69**, 56–67.

Passmore, R. & Draper, M. H. (1965). Energy metabolism. In *Newer Methods of Nutritional Biochemistry*, ed. A. A. Albanese, New York: Academic Press.

Piers, L. S., Soares, M. J. & Shetty, P. S. (1992). Thermic effect of a meal: 2. Role in chronic undernutrition. *British Journal of Nutrition*, **67**, 177–85.

Piers, L. S., Diggavi, S. N., Thangam, S., Van Raaij, J. M. A., Shetty, P. S. & Hautvast, J. G. A. J. (1995). Changes in energy expenditure, anthropometry, and energy intake during the course of pregnancy and lactation in well-nourished Indian women. *American Journal of Clinical Nutrition*, **61**, 501–13.

Poppitt, S. D., Prentice, A. M., Goldberg, G. R. & Whitehead, R. G. (1994). Energy-sparing strategies to protect human fetal growth. *American Journal of Obstetrics and Gynaecology*, **171**, 119–25.

Poppitt, S. D., Prentice, A. M., Jequier, E., Schutz, Y. & Whitehead, R. G. (1993). Evidence of energy sparing in Gambian women during pregnancy: a longitudinal study using whole-body calorimetry. *American Journal of Clinical Nutrition*, **57**, 353–4.

Prentice, A. M., Goldberg, G. R., Davies, H. L., Murgatroyd, P. R. & Scott, W. (1989). Energy sparing adaptations in human pregnancy assessed by whole-body calorimetry. *British Journal of Nutrition*, **62**, 5. 22.

Prentice, A. M. & Whitehead, R. G. (1987). The energetics of human reproduction. *Zoological Society London Symposium*, **57**, 275–384.

Ramanamurthy, P. S. V., Srikantia, S. G. & Gopalan, C. (1962). Energy metabolism in undernourished subjects before and after rehabilitation. *Indian Journal of Medical Research*, **50**, 108–12.

Ravussin, E. & Bogardus, C. (1989). Relationship of genetics, age and physical activity to daily expenditure and fuel utilisation. *American Journal of Clinical Nutrition*, **49**, 968–75.

Rothwell, N. J., Saville, M. E. & Stock, M. J. (1982). Sympathetic and thyroid influences on metabolic rate in fed, fasted and refed rats. *American Journal of Physiology*, **234**, R339–46.

Sherwin, R. & Felig, P. (1987). Starvation in normal humans. In *Endocrinology and Metabolism*. ed. P. Felig, J. D. Baxter, A. E. Broadus and L. A. Froham, pp. 1182–3. New York: McGraw-Hill.

Shetty, P. S. (1980). *Studies on protein and energy restriction and dietary thermogenesis in obesity and chronic undernutrition.* PhD Thesis, University of Cambridge, UK.

Shetty, P. S. (1984). Adaptive changes in basal metabolic rate and lean body mass in chronic undernutrition. *Human Nutrition: Clinical Nutrition*, **38**, 573–81.

Shetty, P. S. (1990). Physiological mechanisms in the adaptive response of metabolic rates to energy restriction. *Nutrition Research Reviews*, 3, 49–74.

Shetty, P. S. (1992). Respiratory quotients and substrate oxidation rates in the fasted and fed state in chronic energy deficiency. In *Protein–Energy interactions*, ed. N. S. Scrimshaw and B. Schurch, pp. 139–50. Switzerland: IDECG.

Shetty, P. S. (1995). Body composition in Malnutrition. In *Body Composition Techniques in Health and Disease*, ed. P. S. W. Davies and T. J. Cole, pp. 71–84. Cambridge: Cambridge University Press.

Sims, E. A. H. (1976). Experimental obesity, dietary induced thermogenesis and their clinical implications. *Clinics in Endocrinology and Metabolism*, 5, 377–95.

Sims, E. A. H. (1986). Energy balance in human beings: the problem in plenitude. *Vitamins and Hormones*, **43**, 1–101.

Soares, M. J., Piers, L. S., Shetty, P. S., Robinson, S. Jackson, A. A. & Waterlow, J. C. (1991). Basal metabolic rate, body composition and whole-body protein turnover in Indian men with differing nutritional status. *Clinical Science*, **81**, 419–25.

Soares, M. J. & Shetty, P. S. (1991). Basal metabolic rates and metabolic efficiency in chronic undernutrition. *European Journal of Clinical Nutrition*, **45**, 363–73.

Srikantia, S. G. (1985). Nutritional adaptation in man. *Proceedings of the Nutrition Society of India*, **31**, 1–16.

Sukhatme, P. V. & Margen, S. (1982). Autoregulatory homeostatic nature of energy balance. *American Journal of Clinical Nutrition*, **35**, 355–65.

Taylor, H. L. & Keys, A. (1950). Adaptation to caloric restriction. *Science*, **112**, 215–18.

Van Raaij, J. M. A., Vermaat-Miedma, S. H., Schonk, C. M., Peck, M. E. M., Hautvast, J. G. A. J. (1987). Energy requirements of pregnancy in The

Netherlands. *Lancet*, **ii**, 953–5.

Venkatachalam, P. S., Srikantia, S. G. & Gopalan, C. (1954). Basal metabolism in nutritional edema. *Metabolism*, **2**, 138–41.

Waterlow, J. C. (1986). Metabolic adaptation to low intakes of energy & protein. *Annual Review of Nutrition*, **6**, 495–526.

Weinsier, R. L., Schutz, Y. & Braco, D. (1992). Re-examination of the relationship of resting metabolic rate to fat-free mass and to the metabolically active components of fat-free mass in humans. *American Journal of Clinical Nutrition*, **55**, 790–4.

Widdowson, E. M. (1962). Nutritional individuality. *Proceedings of the Nutrition Society*, **21**, 121–8.

Widdowson, E. M. (1985). Responses to deficits of dietary energy. In *Nutritional Adaptation in Man.* ed. K. L. Blaxter and J. C. Waterlow, pp. 97–104. London: John Libbey.

9 Possible adaptive mechanisms for energy saving during physical activity

M. RIEU

There is considerable interest in knowing the possible mechanisms by which undernourished individuals are able to reduce their energy expenditure in response to limited energy intake. This knowledge is especially pertinent because about 20–25% of the population of the world are living in less-developing countries where food shortages exist. Many studies have attempted to evaluate the energy intake, the energy output and the work capacity of such populations in order to determine whether energy balance is maintained, and if so, by what adaptive mechanisms.

Several methods have been used but most of them are difficult to apply under field conditions. For instance, energy intake studies are usually completed by means of questionnaires or diet diaries filled in by the subjects themselves with the responses being checked by a nutritionist (Marr, 1971; Boisvert et al., 1988). Apart from problems of literacy which may make completion of forms difficult, accurate knowledge of the chemical composition and energetic equivalent of the ingested food must be known. This is not always easily obtainable in developing countries where the food is often very specific and original (Toury, Giorgi & Favier, 1967; FAO, 1968, 1972; Benefice et al., 1985; Rosetta, 1988).

In order to measure energy output, researchers have often had to rely on responses to questionnaires or diaries which record the amount and extent of physical activities of subjects (Bouchard et al., 1983). Many other methods are used too. For example, the continuous recording of heart rate (HR) throughout the day, can, knowing the linear relationship between HR and the oxygen consumption, be used to determine energy output (Bradfield, 1971; Rieu et al., 1980). Other methods include the recording of the movements of individuals by means of an actometer (Avons et al., 1988); strict surveying and noting by the investigator of the kind and duration of the physical activities performed by subjects (Paffenbarger et al., 1993) or the use of the doubly labelled water, a very elegant but very expensive technique (Coward *et al.*, 1984; Livingstone *et al.*, 1990; Westerterp *et al.*, 1986).

With regard to physical activities in humans, the mechanical work capacity is dependent on the total quantity of metabolic energy (E) that an individual is able to use in a given period of time and on the mechanical efficiency (η):

$$W = E.\eta$$

From a physiological point of view and for a submaximal exercise:

$$E = k(F.\dot{V}O_{2max}).t$$

where k is an unit correspondence factor and F is the maximum fraction of $\dot{V}O_{2max}$ which can be used during the period of exercise (t). An exponential inverse relationship exists between $F\dot{V}O_{2max}$ and t such that the smaller $F.\dot{V}O_{2max}$, the greater is t. This relationship is expressed by the endurance concept which is an individual characteristic (Gleser & Vogel, 1973). For example, one elite male marathon runner can run with $F = 0.9$, whereas an ordinary runner will not be able to perform such a trial with F greater than 0.75. On the other hand, in normal everyday life, if one is working for eight hours, F will not be higher than 0.30–0.35 (Michael, Hutton & Horvath, 1961; Spurr, Barac-Nieto & Maksud, 1975). Thus, in order to evaluate how long a subject will be able to work, it is necessary to know (a) the metabolic power corresponding to a given task and (b) the maximal aerobic power (MAP, expressed in terms of $\dot{V}O_{2max}$) and (c) the endurance level of an individual.

The determination of $\dot{V}O_{2max}$ and of endurance have been made in the laboratory under rigorous conditions which are difficult to reproduce in less-developing countries. Furthermore, the knowledge of the energy cost (C) of each task is necessary in most of previously mentioned methodological approaches which estimate the energy output. In fact, many data banks exist detailing the energy cost of the main current human activities (Passmore & Durnin, 1955; Durnin & Passmore, 1967; Bannister & Brown, 1968; Edholm *et al.*, 1970; Ainsworth *et al.*, 1993). Unfortunately the majority of data collected concerns Western populations rather than third-world people. It is evident that many of the tasks are not similar because most westerners are living an urban lifestyle while many in the third world live in a rural environment (Durnin & Ferro-Luzi, 1982).

To summarize, the potential productivity of humans results from three factors: $\dot{V}O_{2max}$, endurance and the energy cost of the activities. However, the level of energy intake can influence the work capacity and some metabolic effects of malnutrition have been well elucidated. For example, malnutrition is known to reduce the MAP (Barac-Nieto *et al.*, 1980; Spurr, 1988) and fasting to result in diminution of the basal metabolic rate (BMR) (Benedict, 1915).

Obviously the lower MAP found in undernourished populations leads to a reduction in the quantity of work per day a subject is able to produce. Indeed, even assuming that endurance (expressed as maximal percentage of $\dot{V}O_{2max}$ which can be used for a given time) does not change, the absolute quantity of available energy will be lowered. On the other hand, undernourished subjects should take advantage of the reduction in their BMR. Indeed, the whole energy expended during the achievement of an activity is equal to the sum of BMR and of the metabolic power linked to the given task by its duration. Of course, the lower the BMR, the lower will be the total energy cost of a physical trial. But in fact, the problem is more complex because of the drop of BMR would result from two adaptive ways; the first is direct and has been ascribed to an increase in metabolic efficiency, the second which is indirect, would be linked to an anthropometric adaptation, namely the reduced body dimensions (height and weight). But some people have suggested that the smaller the body size, the greater is the energy cost of a given physical activity. For instance, the latter assumption is true when children are compared to adults with regard to the energy cost of running (Cavagna, Fraazetti & Fuchimoto, 1983; Krahenbuhl & Williams, 1992) because such differences disappear when the values of C are expressed taking body size into account. Thus it is possible to suggest that an increase in C would be able to balance the reduction in BMR. Furthermore, it is not proved definitely that chronic malnutrition, as is the case in some less-developed countries, actually results in a decrease in BMR.

The decrease in energy cost of physical activities could be another kind of adaptation that would result in reduction in energy expenditure for the same mechanical work production. In fact, a few studies exist showing that the energy cost of walking (with or without an extra load) and of running, would be lower in people living in developing countries than in Caucasian groups living in western cities (Maloiy *et al.*, 1986; Ferretti *et al.*, 1991). But, to date, it is unknown if such differences result from chance observations or from actual bio-behavioural adaptations.

Energy cost concept

Before considering the several possibilities of adaptation, it is important to exactly define the energy cost concept.

In general, in all tasks:

Total metabolic energy expenditure (E) = metabolic power $(\dot{E})$ × duration of the work (t) and the mechanical efficiency (η) = mechanical power (P)/metabolic power $(\dot{E})$.

From another point of view, the inverse of the mechanical efficiency (η) is the energy cost:

$$C = \frac{\dot{E}}{P} = \frac{1}{\eta}$$

It follows that the higher η is, the lower C will be.

But as it is very difficult to measure with accuracy the actual mechanical work of most of tasks, a more global approach is usually taken by determining the energy expenditure necessary to perform a given task. This task must be clearly defined, and the time between the start and the finish of the work carefully noted.

As an example, the determination of the energy cost of walking and running and the analysis of the factors which can alter them will be briefly described. This model can be more or less easily adapted to the studies of other physical activity modes.

Determination and significance of the energy cost of running and walking

The energy cost $(C$ = unit of energy expenditure per unit of distance$)$ of running and walking correspond to the ratio between the metabolic power $(\dot{E}$ = metabolic energy expenditure per unit of time$)$ and the running or walking speed (v). At a given speed, C depends on the air resistance (aerodynamic force = E_a), the gradient of the slope (E_g) and many other individual and environmental factors (E_s) (from di Prampero, 1986).

$$C = E_s + E_a + E_g$$

The quantity of energy wasted per unit of distance against the aerodynamic force (E_a) is proportional to the square of the running speed but can be neglected when this is less than about 5 m s^{-1}:

$$E_a = K.v^2 \cdot \eta^{-1}$$

where K is a constant which depends on the body area projected on the frontal plane and on the air density which is itself dependent on altitude and on the mechanical efficiency of the running or of the walking

($\eta = 0.20$–0.25 and $K.\eta \sim 0.40$ in running). However, according to its direction, the wind can either increase or decrease in E_a.

When walking or running uphill, the extra energy (E_g) per unit of distance required to exceed the force of gravity is given by the formula:

$$E_g = M.S/100 \cdot g \cdot \eta^{-1}$$

where E_g is the product of the body mass (M) by the slope (S in %) and by the acceleration of gravity (g$=9.81$ m s^{-2} and g $\eta^{-1} \sim 39.2$).

The results are expressed in the MKS system in terms of joule per metre (J. m^{-1}).

The energy cost of walking and running have a minimum at a slope of about -10%, increasing on both sides of this curve which is U shaped (Margaria *et al.*, 1963).

Obviously, when the subject is running on a treadmill, E_a is nil and if the ground is flat, $E_g = 0$ and $C = E_s = \dot{E}_s/v$ where $\dot{E}_s$ is the metabolic power used against the non-aerodynamic and gravity forces. It is measured in the laboratory under standard conditions by means of oxygen consumption determination.

The calculation of the value of the standard energy cost (C_s) is achieved when the intensity of the work remains below the anaerobic threshold and when the steady-state oxygen consumption ($\dot{V}O_{2w}$) has been reached. Under these conditions, the anaerobic energy sources are supposed to be null.

In practice, the subject runs or walks on a level treadmill. The speed of the ergometer must remain constant and below the level defined in the Wasserman methods (Wasserman *et al.*, 1973) as being the anaerobic threshold. The duration of the test is about 7–10 minutes. Under aerobic and steady-state conditions, the standard energy cost is equal to the oxygen cost:

$$C_s = \dot{V}O_{2w} \times 60 / v$$

where 60 is a correspondence factor with time

$\dot{V}O_{2w}$ is expressed in ml O_2 kg^{-1} min^{-1}; v in km h^{-1} and C_s in ml O_2 kg^{-1} km^{-1};

and the net energy cost:

$$C_{s\ net} = (\dot{V}O_{2w} - \dot{V}O_{2\ rest}) \times 60 / v$$

The results reported in the literature show great inter-individual variability. Effectively, the C_s range for running is 165–230 ml O_2 kg^{-1} km^{-1} or

3.5–4.9 J kg^{-1} m^{-1} which is broadly equivalent to 1 Cal kg^{-1} km^{-1} (Brükner, 1986).

Under these standard conditions energy expenditure results from several factors:

(i)   Contractions of locomotor muscles at each stride lead to changes in the kinetic energy due to acceleration (propulsion phase:positive work) and deceleration (reception phase:negative work) of the body mass centre.

(ii)   The friction of the articular and/or aponevrotic surfaces between themselves and the one of the tendons slipping in their bony groove;

(iii)   The muscular contractions associated but not directly involved in the change of position of the body mass centre, i.e. muscular work for the maintenance of posture and equilibrium as the movements of arms; contractions of the respiratory muscles and of the heart.

Standard energy cost alterations in walking and running

Non-environmental factors

Many authors have studied the relationship between the speed of running or walking and C_s.

In running, most studies show that C_s is independent of the speed without limits. (Margaria *et al.*, 1963; Dill, 1965; van der Walt & Windham, 1973; Hagberg & Coyle, 1984).

In walking, the relationship between C_s and speed is very different from running. At first, when the speed is slow, C_s is relatively high; then it decreases as the speed increases to reach a constant value (about half that for running) between 1 and 1.4 m s^{-1}; afterwards C_s increases again at higher speeds (Bobbert, 1960; Cotes & Meade, 1960; Dill, 1965). Above a speed of about 8 km h^{-1}, running is more economical than walking and the subjects usually spontaneously change their type of locomotion from one to another (Grillner *et al.*, 1979; Thorstensson & Roberthson, 1987).

There are a few papers relating the effects of fatigue due to running on C_s (Williams, Snow & Agruss, 1988; Morgan *et al.*, 1990). In one of our studies (Brükner *et al.*, 1991), we determined the C_s net of ten amateur runners before and immediately after running 15, 32 or 42 km on an indoor track at a constant speed. The C_s was measured on a treadmill at the same speed, and each run was performed twice. Our results show that C_s only seems to be affected to a minor extent by fatigue. So, in these experimental conditions, the increase in C_s relative to distance was small, amounting on average to 0.08% km^{-1} (0.0029 J kg^{-1} m^{-1}).

The influence of anthropometric characteristics have been studied by many authors. Since body weight is dependent on body size and on the percentage of fat, it is very difficult to determine the respective roles of each of these factors which are also associated with age and sex. However, when the running energy cost is not referred to the body weight (i.e. expressed in ml O_2 km^{-1}), it is strongly correlated with all these parameters.

On the other hand, very few papers report on C_s expressed per unit of body mass (J or ml O_2 km^{-1} kg^{-1}) and those that do, show conflicting results. One author found a negative correlation between C_s and body weight (Williams & Cavanagh, 1987). But in most reports, C_s and body weight are considered as independent values (Morgan, Martin & Krahenbuhl, 1989; Martin & Morgan, 1992).

However there is an important difference between C_s referring to total body weight and C_s in relation to lean body mass. For example, let us compare two subjects A and B, weighing 60 and 80 kg, with a fat ratio of 6% and 11%, respectively. Accordingly, subject A will have a C_s of 187.5 ml O_2 kg^{-1} km^{-1} while subject B will have a C_s of 168.5 ml O_2 kg^{-1} km^{-1}. When taking the respective lean body mass into consideration, the recalculated C_s becomes 199.4 and 205.8 ml O_2 kg^{-1} km^{-1} for subjects A and B, respectively. These modified values reflect more accurately the energetic expenditure of the active muscle mass.

In adults, there is no correlation between C_s and height (Pate *et al.*, 1992; Bourdin *et al.*, 1993). Nevertheless, one study showed a significant negative correlation between the running C_s and the length of the legs as well as for length of the feet (Williams & Cavanagh, 1987). When running, each subject spontaneously adopts the most economic stride length for themselves and, at a given running speed, any alteration results in an increase of C_s (Hogberg, 1952; Knuttgen, 1961; Menier & Pugh, 1968; Morgan & Martin, 1986). Usually, the quicker the running speed, the longer is the stride. This is true until a given speed limit is reached. Only by increasing the stride frequency can further increase in running speed be achieved (Cavanagh & Kram, 1989).

No convincing explanation exists about the optimal spontaneous adaptation of the stride length to the running speed. Analysis of the ratio between the stride length (*SI*) and the body height (*H*) or the leg size (*L*) have resulted in very contradictory results. However Alexander's hypothesis which attempt to link the *SI/L* ratio to the running speed is attractive but has yet to be confirmed (Alexander, 1984).

C_s also seems to depend on the type of muscular fibres. In those of the slow type, the mechanical efficiency should be better at low contraction speed than at higher speed and the opposite for the fast type fibres (Wendt & Gibbs, 1974; Goldspink, 1978; Stuart *et al.*, 1981). The results cited in these papers can account for the differences in efficiency observed between the long distance runners and sprinters according to running speed (Kaneko, 1990).

Environmental factors

The nature of the ground is of considerable importance since the C_s value will obviously be lower when a subject is running on an indoor tartan track than when s/he is floundering in mud. For example, athletic performances broadly depend on the mechanical characteristics of the running track surface (Soule & Goldman, 1972). In human locomotion, equipment is often intermediate between the body and the terrain. This is the case not only in cycling, skiing, ice-skating but also in walking and running where the mechanical characteristics of the sole of shoes as elasticity or absorption capacity of shock can alter the energy cost of running or of walking.

But more important is the weight of the shoes and more generally the influence of all extra loads on the energy cost (Soule & Goldman, 1969; Pandolf, Givoni & Goldman, 1977; Cavanagh & Kram, 1985; Martin, 1985; Pate *et al.*, 1992). Indeed, many studies have shown that the metabolic energy expenditure is lower if the extra weight is supported by the trunk than if it is carried on the body extremities such as the hands or the feet. In the first condition, the increase in $\dot{V}O_2$ will be only 0.1% per 100 g of extra weight; in the second one, $\dot{V}O_2$ may increase until 1% per 100 g of extra weight (Owens, Al-Ahmed & Moffatt, 1989). In addition, when a person is carrying an extra weight in a rucksack, the posture changes and the trunk moves in front of the vertical plane.

The influence of the thermal environment on the energy cost of physical work is unclear. Indeed, many authors have shown a positive correlation between $\dot{V}O_2$ during prolonged submaximal exercise performed in hyperthermic environment and body temperature (Consolazio *et al.*, 1963; Durnin & Haisman, 1966). On the other hand, in other studies increasing rectal temperature had no effect on $\dot{V}O_2$ which could even decrease. One hypothesis that has been put forward is that the mechanical efficiency of muscles increases when their temperature is elevated (Asmussen & Böje, 1945; Rowell *et al.*, 1969).

Energy cost adaptation in walking and running

Several studies have estimated the energy expenditure of specific activities in third-world populations by means of oxygen consumption measured during the period when the subjects were completing their task (Panter-Brick, 1992). The values were converted into kcal min^{-1} or into watts. The results, therefore can be interpreted as energy output. The author observed that the energy output of the studied people was lower than was expected. However, the physiological significance of this result is uncertain because it might indicate that the subjects were working slowly. If this were the case, then it is more appropriate to call it a behavioural adaptation rather than a biological adaptation. A true biological adaptation is one where one subject has lower energy expenditure than another subject while performing the same quantity of work, because the mechanical efficiency (η) is higher. If there is not any increase in efficiency, the decrease in energy output results in an increase in work time, and finally the same quantity of energy would be expended. Nevertheless, knowing if there is any negative relationship between η and mechanical power is of some importance (Gaesser & Brooks, 1975). It is known that, in walking, η decreases while the speed increases; but this is not the case for running, where the efficiency is independent of speed.

The question is more complicated when the very complex tasks seen in field work are considered. In order to give an answer to this problem, it would be necessary to complete several studies about each specific task.

A second question which can be asked is if a population is able to perform a given task at a certain speed in a more economical way than another population, is that the consequence of adaptation? To our knowledge, only two studies actually suggest this possibility.

The first study was completed in East Africa and women carrying loads on the top of their heads were studied in a laboratory setting (Maloiy *et al.*, 1986). The authors showed that (a) when the load does not exceed 20% of body weight, there is no increase in the rate of oxygen consumption at a given walking speed. Above this relative load value, $\dot{V}O_2$ linearly increased in direct proportion to the load (b) for all extra load values, the energy output was lower in East African women than in a Caucasian control population.

The study can be criticized because (a) the number of women studied was very small (five only) and no reliable statistical analysis could be carried out on the data, (b) the relationship between oxygen consumption rate and relative extra loads involves extrapolation of the regression line and it could well be that this relationship is curvilinear, but with only 5 subjects it is not possible to say, (c) the comparison with the Caucasian population was not reliable because the latter comprised young male army recruits carrying a load in a backpack. Moreover, the measurements were performed in a different environmental setting ten years earlier by a different research team using other methods (Pandolf *et al.*, 1977).

In the second study, the energy cost of walking and running were determined in a laboratory for African pygmies (Ferretti *et al.*, 1991). The authors showed that the net value for running C_s was lower in pygmies than in elite Caucasian endurance runners but there was no difference in walking.

In reality, it is very difficult to say whether or not such studies provide substantive evidence for a biological adaptation. If the energy cost of a task is compared between local people and Caucasians, the technical aspect will be all the more important, as the activity will be specific and complex. For instance, it would be unreasonable to compare the energy cost of carrying loads on the head between women using this carriage method since their childhood and subjects which have only recently learned this kind of exercise. Can a subject decrease his whole energy expenditure to achieve a given task (with the exception of technical improvements) ? and, if so, is such an adaptation biological or cultural?

In sport, the physical training of long distance runners represents an interesting physiological model, because the technical aspects remain secondary although they are not all insignificant. So, the energy cost of running by an elite endurance runner is lower than one of a sedentary person (Bransford & Howley, 1977; Dolgener, 1982). Many hypotheses have been put forward to explain this observation. They include:

(i)   a more efficient cardio-respiratory system in the endurance training subjects (Clausen, 1977; Blomquist & Saltin, 1983);

(ii)   the influence of extreme percentages of slow twitch (ST) fibres in the muscles of the endurance athletes (Costill, Fink & Pollock, 1976; Saltin *et al.*, 1977), the mechanical efficiency of these being higher at slow contraction rate (Suzuki, 1979). That may be a consequence of the different elastic behaviour of the slow twitch and of the fast twitch fibres, the former being able to re-use greater amounts of stored energy than the latter (Bosco *et al.*, 1986; Morgan & Craib, 1992);

(iii)   anthropometric characteristics which take into account the size of the subject and the length of the legs and the strides.

Unfortunately, the longitudinal studies have not determined the important factors resulting from endurance training except for the undeniable improvement of the efficiency of the oxygen transport system and the obvious increase in the oxidative capacity of all the fibre types of the trained muscles (Holloszy & Booth, 1976). The possibility that the FT fibres can change into ST fibres is still disputed in spite of the fact that the change of the myosin isoforms from the slow into the fast forms is now very well established (Howald, 1982; Baumann *et al.*, 1987; Booth & Thomason, 1991). Consequently there is interrogation with regard to an eventual increase in the re-use of elastic energy stored in the muscles involved in running. On the other hand, no correlation has been established between the modifications of the gait (length and frequency of strides) or of the vertical displacements of the body mass centre with energy cost changes. Nevertheless, cross-sectional studies have shown that, according to the speed, the increase in the extensiveness of the angular movements of the thigh, of the knee and of the ankle was greater in athletes than in untrained people (Hoshikawa, Matsui & Miyashita, 1973). But the longitudinal studies have given disappointing and contradictory results probably because the training programmes (intermittent or continuous) and the experimental subjects in respect to age and athletic level were too dissimilar.

However, a very recent study has provided new information (Brisswalter, 1994). This study was performed on 14 middle distance runners. The subjects were asked to change their training in quality (from intermittent to continuous mode) and intensity (from supramaximal to submaximal power $= 80\%$ $\dot{V}O_2$ max) and in quantity (from 42 km to 120 km a week). The C_s determination was done at the speed of training and many physiological and kinematic parameters were recorded before and after a training period of three weeks. After training, subjects showed a decrease of 10% in C_s and associated with this was a decrease of 10% in ventilatory gas output and a reduction of 25% in the vertical oscillations of the centre of gravity of the body. All the subjects had changed their stride length, half of them showing an increase while the others showed a decrease.

These results suggest that training can modify the C_s in two ways: (i) increasing the efficiency of the oxygen transport system, and (ii) alterations in the biomechanics of running leading to decreased potential energy changes.

It should be noted that aerobic conditions are the best for performing physical work. So, if anaerobic energy sources are involved, then metabolic energy expenditure will increase. Indeed, the withdrawal of accumulated lactate during exercise will result in extra oxygen consumption during the recovery period which will have to be included in the determination of the energy cost of the given task. In fact, the endurance training can increase the aerobic capacity of the subject and in consequence lead to a decrease of the energy cost of the work.

What about the rural third-world people? The problem remains: is their net energy cost of a given task lower than that of urban Western populations? In order to answer this question, it would be necessary to perform studies with a large number of well selected subjects. These people would be asked to undertake in a laboratory a very simple task requiring little or no practice; the task would have to be easily repeatable, involving for a given time a constant and verifiable mechanical power in strictly aerobic conditions. The measurement of total oxygen consumption during the work and the recovery period and the comparison of the data with control subjects could allow definitive conclusions to be obtained. In the case of positive results, i.e. higher mechanical efficiency in experimental population than in controls, it would be necessary to undertake further research about the cardiac efficiency at rest and during exercise, metabolic and hormonal behaviour, the macro/microscopical and molecular characteristics of muscles, the anthropometrical and the biomechanical pattern of locomotion.

Several questions need to be answered in the near future. For instance, what is the influence of the body fat distribution on the energy cost taking into consideration that an extra load carried on the lower extremities results in greater energy expenditure than when carried on the trunk? It would be very interesting to develop our knowledge about the thyroid status in the investigated populations. Indeed, there is some experimental work which suggests that hyperthyroidism would involve a shift from slow to fast myosin isoforms and would help the conversion of ST into FT fibres. The thyroid deficiency would give opposite results (Butler-Browne, Herlicoviez & Whalen, 1984; Fitzsimons, Herrick & Baldwin, 1990; Ianuzzo, Hamilton & Lee, 1991). Thus, the mechanical efficiency of muscle seems in part to depend on its fibre type composition.

In conclusion, the reduction of energy expenditure through a combination of limited energy intake and to the environmental conditions can result from two distinct mechanisms corresponding to either a behavioural adaptation or to a physiological one (Fig. 9.1). Subjects reduce their energy output by means of a decrease in the whole quantity of mechanical work performed each day. In the second case, the whole quantity of work does not change but the energy output is reduced because of an increase in the mechanical efficiency namely a decrease in the energy cost of physical tasks, and many factors are able to modify the mechanical efficiency of physical work (Fig. 9.2). Independently of current environmental factors, there are two ways in which the latter may occur: through technical improvement in

Mode	t	P	η	C	BMR	$\dot{E}$	p	E
(i) Physiological	=	=	↑	↓	= or ↓	↓	=	↓
(ii) Behavioural	=	↓	=	=	=	↓	↓↓	↓
(iii) Both	=	↓	↑	↓	= or ↓	↓	= or ↓	↓

Fig. 9.1. Alteration of the metabolic energy expenditure and of the productivity related to the different patterns of change of mechanical efficiency and tasks energy cost.

Mode = adaption mode
  t = mechanical work time = duration of the work
  P = mechanical work power (work rate) (watts)
  E = total metabolic expenditure (joules)
  $\dot{E}$ = metabolic power = E per unit of time
  η = mechanical efficiency = $P/\dot{E}$ = number without dimension. In human
    ~ 0.020–0.25
  C = energy cost = $1/\eta$ = E per unit of work and in a wider sense = E
    necessary to perform a given task. For instance in locomotion: joules
    per metre = J m^{-1}
BMR = basal metabolic rate
  p = productivity = $P \times t$

Under the situation (iii), η increases because P decreases, assuming that negative correlation $[\eta = -f(P)]$ exists between work power and efficiency. (It is the case in walking, but not in running.)

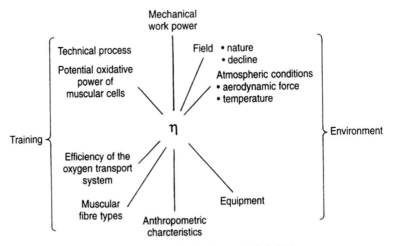

Fig. 9.2. Factors likely to modify the mechanical efficiency.

performing the task or by a biological adaptation although the latter has still to be proved at population level and the eventual mechanisms remain speculative.

Another problem is to know if, in the very well-adapted populations, the increase in energy intake will result in an improvement in productivity namely of the energy output. To date, no definite answer has been made to this question.

REFERENCES

Ainsworth, B. E., Haskell, W. L., Leon, A. S., Jacobs, J. R., D. R., Montoye, H. J., Sallis, J. F. & Paffenbarger, JR., R. S. (1993). Compendium of Physical Activities: classification of energy costs of human physical activities. *Medicine and Science in Sports and Exercise*, **25**, 71–80.

Alexander, R. M. (1984). Stride length and speed adults, children and fossils hominids. *American Journal of Physiological Anthropology*, **63**, 23–7.

Alexander, R. M. (1992). Comparative aspects of human activity. In *Physical Activity and Health*, SSHB Symposium 34, ed. N. G. Norgan, pp. 7–19.

Asmussen, E. & Böje, O. (1945). Body temperature and capacity for work. *Acta Physiologica Scandinavica*, **10**, 1–22.

Avons, P., Garthwaite, P., Davies, H. & Murgatroyd, J. W. (1988). Approaches to estimating activity in the community: calorimetric validation of actometers and heart rate monitoring. *European Journal of Clinical Nutrition*, **42**, 185–96.

Bannister, E. W. & Brown, S. R. (1968). The relative requirements of physical activity. In *Exercise Physiology*, ed. H. B. Falls, New York: Academic Press.

Barac-Nieto, M., Spurr, G. B., Dahners, H. W. & Maksud, M. G. (1980). Aerobic work capacity and endurance during nutritional repletion of severely undernourished men. *American Journal of Clinical Nutrition*, **33**, 2268–75.

Baumann, H., Jäggi, M., Soland, F., Howald, H. & Schaub, M. C. (1987). Exercise training induces transitions of myosin isoform subunits within histochemically typed human muscle fibres. *European Journal of Physiology*, **409**, 349–60.

Benedict F. G. (1915). A study of prolonged fasting. *Carnegie Institute*, **203**, p. 416. Washington DC.

Benefice, E., Simondon, F., Chevassus-Agnès, S. & Ndiaye, A. M. (1985). Etudes de nutrition dans la moyenne vallée du Sénégal. 1. Evolution de la consommation alimentaire depuis 1958 et structure actuelle de la ration. *Bulletin de la Société de Pathologie Exotique*, **78**, 110–18.

Blomquist, C. G. & Saltin, B. (1983). Cardiovascular adaptation to physical training. *Annual Review of Physiology*, **45**, 169–90.

Bobbert, A. C. (1960). Energy expenditure in level and grade walking. *Journal of Applied Physiology*, **15**, 1015–21.

Boisvert, P., Washburn, R. A., Montoye, H. J. & Léger, L. (1988). Mesure et évaluation de l'activité physique par questionnaire. Questionnaires utilisés dans la littérature anglo-saxonne. *Science and Sports*, **3**, 245–62.

Booth, F. W. & Thomason, D. B. (1991). Molecular and cellular adaptation of muscle in response to exercise: perspectives of various models. *Physiological Reviews*, **71**, 541–85.

Bosco, B., Tihanyi, J., Latteri, F., Fekete, G., Apor, P. & Rusko, H. (1986). The effect of fatigue on store and re-use of elastic energy in slow and fast types of human skeletal muscle. *Acta Physiologica Scandinavica*, **128**, 109–17.

Bouchard, C., Leblanc, C., Lortie, G., Savard, R. & Theriault, G. (1983). A method to assess energy expenditure in children and adults. *American Journal of Clinical Nutrition*, **37**, 461–7.

Bourdin, M., Pasthne, J., Germain, M. & Lacour, J. R. (1993). Influence of training, sex, age and body mass on the energy cost of running. *European Journal of Applied Physiology*, **66**, 439–44.

Bradfield, R. (1971). A technique for determination of usual daily energy expenditure in the field. *American Journal of Clinical Nutrition*, **24**, 1148–54.

Bransford, D. R. & Howley, E. T. (1977). Oxygen cost of running in trained and untrained men and women. *Medicine, Science and Sports*, **9**, 41–4.

Brisswalter, J. (1994). Processus d'adaptation à la course à pied pour des intensites sous-maximales. Thèse de doctorat. Université Paris V.

Brükner, J. J. (1986). Le coût énergétique de la course d'endurance. Thèse présentée à la faculté de médecine de l'Université de Genève, n°7023, p. 63. Genève Editions Médecine et Hygiène.

Brükner, J. C., Atchou, G., Capelli, C., Duvallet, A., Barrault, D., Joussellin, E., Rieu, M. & di Prampero, P. E. (1991). The energy cost of running increases with the distance covered. *European Journal of Applied Physiology*, **62**, 385–9.

Butler-Browne, G. S., Herlicoviez, D. & Whalen, R. G. (1984). Effects of hypo-thyroidism on myosin isozyme transitions in developing rat muscle. *Federation of European Biochemical Societies*, **166**, 71–5.

Cavagna, G. A., Franzetti, P. & Fuchimoto, T. (1983). The mechanics of walking in children. *Journal of Physiology (London)*, **343**, 323–39.

Cavanagh, P. R. & Kram, R. (1985). Mechanical and muscular factors affecting the efficiency of human movement. *Medicine and Science in Sports and Exercise*, **17**, 326–31.

Cavanagh, P. R. & Kram, R. (1989). Stride length in distance running: velocity, body dimensions, and added mass effects. *Medicine and Science in Sports and Exercise*, **21**, 467–79.

Clausen, J. P. (1977). Effect of physical training on cardio-vascular adjustments to exercise in man. *Physiological Reviews*, **57**, 779–815.

Consolazio, C. F., Matoush, L. O., Nelson, R. A., Torres, J. B. & Isaac, G. J. (1963). Environmental temperature and energy expenditure. *Journal of Applied Physiology*, **18**, 65–8.

Costill, D. L., Fink, W. J. & Pollock, M. L. (1976). Muscle fibre composition and enzyme activities of elite distance runners. *Medicine, Science and Sports*, **8**, 96–100.

Cotes, J. E. & Meade, F. (1960). The energy expenditure and mechanical energy demand in walking. *Ergonomics*, **3**, 97–119.

Coward, W. A., Prentice, A. M., Mugatroyd, P. R., Davies H. L., Cole, T. J., Sawyer, M., Goldberg, G. R., Halliday, D. & McNamara, J. P. (1984). Measurement of CO_2 and water production in man using ^2H, ^{18}O-labelled H_2O; comparisons between calorimeter and isotope values. In *Human Energy*

Metabolism, n°5, 126–8, Wageningen. Edited by A. J. A. van Es European Nutrition Report.

Dill, D. B. (1965). Oxygen used in horizontal and grade walking and running on the treadmill. *Journal of Applied Physiology*, **20**, 19–22.

di Prampero, P. E. (1986). The energy cost of human locomotion on land and in water. *International Journal of Sports Medicine*, **7**, 55–72.

Dolgener, F. (1982). Oxygen cost of walking and running in untrained, sprint trained, endurance trained females. *Journal of Sports Medicine and Physical Fitness*, **22**, 60–5.

Durnin, J. V. G. A. & Ferro-Luzzi, A. (1982). Conducting and reporting studies on human energy intake and output: suggested standards. *American Journal of Clinical Nutrition*, **35**, 624–6.

Durnin, J. V. G. A. & Haisman, M. F. (1966). The effects of hot environments on the energy metabolism of men doing standardized physical work. *Journal of Physiology (London)*, **183**, 75P.

Durnin, J. V. G. A. & Passmore, R. (1967). *Energy, Work and Leisure*. London: Heinemann Educational Books.

Edholm, O. G., Adam, J. M., Healy, M. J. R., Wolff, H. S., Goldsmith, R. & Best, T. W. (1970). Food intake and energy expenditure of army recruits. *British Journal of Nutrition*, **24**, 1091–107.

FAO (1968). Tables de composition alimentaire pour l'Afrique. *Food and Agriculture Organization*, Rome.

FAO (1972). Tables de composition alimentaire pour l'Asie du Sud-Est. *Food and Agriculture Organization*, Rome.

Ferretti, G., Atchou, G., Grassi, B., Marconi, C. & Cerretelli, P. (1991). Energetics of locomotion in African pygmies. *European Journal of Applied Physiology*, **62**, 7–10.

Fitsimons, D. P., Herrick, R. E. & Baldwin, K. M. (1990). Isomyosin distributions in rodent muscles: effects of altered thyroid state. *Journal of Applied Physiology*, **69**, 321–7.

Gaesser, G. A. & Brooks, G. A. (1975). Muscular efficiency during steady-state exercise: effects of speed and work rate. *Journal of Applied Physiology*, **38**, 1132–9.

Gleser, M. A. & Vogel, J. A. (1973). Endurance capacity for prolonged exercise on the bicycle ergometer. *Journal of Applied Physiology*, **34**, 438–42.

Goldspink, G. (1978). Energy turnover during contraction of different types of muscle. In *Biomechanics VI-A*, ed. E. Asmussen and K. Jorgensen, pp. 27–39. Baltimore MD: University Park Press.

Grillner, S., Halberstma, J., Nilsson, J. & Thorstensson, A. (1979). The adaptation to speed in human locomotion. *Brain Research*, **165**, 177–82.

Hagberg, J. M. & Coyle, E. F. (1984). Physiological comparison of competitive race walking and running. *International Journal of Sports Medicine*, **5**, 74–7.

Hogberg, P. (1952). How do stride length and stride frequency influence energy output in running. *International Zeitung Angew Physiology*, **14**, 437.

Holloszy, J. O. & Booth, F. W. (1976). Biochemical adaptations to endurance exercise in muscle. *Annual Review of Physiology*, **38**, 273–91.

Hoshikawa, T., Matsui, H. & Miyashita, M. (1973). Analysis of running pattern in relation to speed. In *Biomechanics III*, 342–8.

Howald, H. (1982). Training-induced morphological and functional changes in skeletal muscle. *International Journal of Sports Medicine*, **3**, 1–12.

Ianuzzo, C. D., Hamilton, N. & Li, B. (1991). Competitive control of myosin expression: hypertrophys vs. hyperthyroidism. *Journal of Applied Physiology*, **70**, 2328–30.

Kaneko, M. (1990). Mechanics and energetics in running with special reference to efficiency. *Journal of Biomechanics*, **23**, 57–63.

Knuttgen, H. G. (1961). Oxygen uptake and pulse rate while running with undetermined and determined stride lengths at different speeds. *Acta Physiologica Scandinavica*, **52**, 366–71.

Krahenbuhl, G. S. & Williams, T. J. (1992). Running economy: changes with age during childhood and adolescence. *Medicine and Science in Sports and Exercise*, **24**, 462–6.

Livingstone, M. B. E., Prentice, A. M., Coward, W. A., Ceesay, S. M., Strain, J. J., McKenna, P. G., Nevin, G. B., Barker, M. E. & Hickey, R. J. (1990). Simultaneous measurement of free-living energy expenditure by the doubly labeled water method and heart-rate monitoring. *American Journal of Clinical Nutrition*, **52**, 59–65.

Maloiy, G. M. O., Heglund, N. C., Prager, L. M., Cavagna, G. A. & Taylor, C. R. (1986). Energetic cost of carrying loads: have African women discovered an economic way? *Nature*, **319**, 668–9.

Margaria, R., Cerretelli, P., Aghemo, P. & Sassi, G. (1963). Energy cost of running. *Journal of Applied Physiology*, **18**, 367–70.

Marr, J. W. (1971). Individual dietary surveys: purposes and methods. *World Review of Nutrition and Dietetic*, **13**, 105–64.

Martin, P. E. (1985). Mechanical and physiological responses to lower extremity loading during running. *Medicine and Science in Sports and Exercise*, **17**, 427–33.

Martin, P. E. & Morgan, D. W. (1992). Biomechanical considerations for economical walking and running. *Medicine and Science in Sports and Exercise*, **24**, 467–74.

Menier, D. R. & Pugh, L. G. C. E. (1968). The relation of oxygen intake and velocity of walking and running in competition walkers. *Journal of Physiology (London)*, **197**, 717–21.

Michael, E. D., Hutton, K. E. & Horvath, S. M. (1961). Cardiorespiratory responses during prolonged exercise. *Journal of Applied Physiology*, **16**, 997–1000.

Morgan, D. W. & Craib, M. (1992). Physiological aspects of running economy. *Medicine and Science in Sports and Exercise*, **24**, 456–61.

Morgan, D. W. & Martin, P. E. (1986). Effects of stride alteration on race walking economy. *Canadian Journal of Applied Sports and Science*, **11**, 211–17.

Morgan, D. W., Martin, P. E., Baldini, F. D. & Krahenbuhl, G. S. (1990). Effects of a prolonged maximal run on running economy and running mechanics. *Medicine and Science in Sports and Exercise*, **22**, 834–40.

Morgan, D. W., Martin, P. E. & Krahenbuhl, G. (1989). Factors affecting running economy. *Sports Medicine*, **7**, 310–30.

Owens, S. G., Al-Ahmed, A. & Moffatt, R. J. (1989). Physiological effects of walking and running with hand-held weights. *Journal of Sports Medicine and Physical Fitness*, **29**, 384–7.

Paffenbarger, R. S., JR., Blair S. N., Lee, I-M. & Hyde, R. T. (1993). Measurement of physical activity to assess health effects in free-living populations. *Medicine and Science in Sports and Exercise*, **25**, 60–70.

Pandolf, K. B., Givoni, B. & Goldman, R. F. (1977). Predicting energy expenditure with loads while standing or walking very slowly. *Journal of Applied Physiology*, **43**, 577–81.

Panter-Brick, C. (1992). The energy cost of common tasks in rural Nepal: levels of energy expenditure compatible with sustained physical activity. *European Journal of Applied Physiology*, **64**, 477–84.

Passmore, R. & Durnin, J. V. G. A. (1955). Human energy expenditure. *Physiological Reviews*, **35**, 801–40.

Pate, R. R., Macera, C. A., Bailey, S. P., Bartoli, W. P. & Powell, K. E. (1992). Physiological, anthropometric, and training correlates of running economy. *Medicine and Science in Sports and Exercise*, **24**, 1128–33.

Rieu, M., Fouillot, J. P., Devars, J. & Cocquerez, J. P. (1980). Automatic analysis of electrocardiogram long term recording with reference to physical activity events analysis. *International Series on Sports Sciences, Children and Exercise IX*, **10**, 183–92. University Park Press.

Rosetta, L. (1988). Seasonal variations in food consumption by Serere families in Senegal. *Ecology of Food and Nutrition*, **20**, 275–86.

Rowell, L. B., Brengelmann, G. L., Murray, J. A., Kraning II, K. K. & Kusumi, F. (1969). Human metabolic responses to hyperthermia during mild to maximal exercise. *Journal of Applied Physiology*, **26**, 395–402.

Saltin, B., Henrikson, J., Nygaard, E. & Andersen, P. (1977). Fiber types and metabolic potentials of skeletal muscles in sedentary man and endurance runners. *Annals New York Academy of Sciences*, **301**, 3–29.

Soule, R. G. & Goldman, R. F. (1969). Energy cost of loads carried on the head, hands, or feet. *Journal of Applied Physiology*, **27**, 687–90.

Soule, R. G. & Goldman, R. F. (1972). Terrain coefficients for energy cost prediction. *Journal of Applied Physiology*, **32**, 706–8.

Spurr, G. B. (1988). Marginal malnutrition in childhood: Implications for adult work capacity and productivity. In *Capacity for Work in the Tropics*, ed. K. J. Collins & D. F. Roberts, pp. 107–40. Cambridge: Cambridge University Press.

Spurr, G. B., Barac-Nieto, M. & Maksud, M. G. (1975). Energy expenditure cutting sugar cane. *Journal of Applied Physiology*, **39**, 990–6.

Stuart, M. K., Howley, E. T., Gladden, L. B. & Cox, R. H. (1981). Efficiency of trained subjects differing in maximal oxygen uptake and type of training. *Journal of Applied Physiology*, **50**, 444–9.

Suzuki, Y. (1979). Mechanical efficiency of fast- and slow-twitch muscle fibers in man during cycling. *Journal of Applied Physiology*, **47**, 263–7.

Thorstensson, A. & Roberthson, H. (1987). Adaptations to changing speed in human locomotion: speed transition between walking and running. *Acta Physiologica Scandinavica*, **131**, 211–14.

Toury, J., Giorgi, R. & Favier, J. C. (1967). Aliments de l'Ouest africain, tables de composition. *Annales de la Nutrition et de l'Alimentation*, **21**, 73–127.

van der Walt, W. H. & Wyndham, C. H. (1973). An equation for prediction of energy expenditure of walking and running. *Journal of Applied Physiology*, **34**, 559–63.

Wasserman, K., Whipp, B. J., Koyal, S. N. & Beaver, W. L. (1973). Anaerobic threshold and respiratory gas exchange during exercise. *Journal of Applied Physiology*, **35**, 236–43.

Wendt, I. R. & Gibbs, O. L. (1974). Energy production of mammalian fast- and slow-twitch muscle during development. *American Journal of Physiology*, **226**, 642–7.

Westerterp, K. R., Saris, W. H. M., van Es, M. & ten Hoor, F. (1986). Use of the doubly-labeled water technique in humans during heavy sustained exercise. *Journal of Applied Physiology*, **61**, 2162–7.

Williams, K. R. & Cavanagh, P. R. (1987). Relationship between distance running mechanics, running economy, and performance. *Journal of Applied Physiology*, **63**, 1236–45.

Williams, K. R., Snow, R. & Agruss, C. (1988). Changes in distance running kinematics with fatigue. *Medicine and Science in Sports and Exercise*, **20**, S49.

10 *Body composition and fertility: methodological considerations*

N. G. NORGAN

Introduction

Measurement of human body composition has become widespread in biological anthropology and in health-related studies. In studying variability in human fertility, investigators have wanted to measure the body composition, the lean:fat ratio; the minimum level of fatness; and the changing energy balance by serial measurements of body composition in adolescent girls and women of reproductive age to determine if these affect fertility. On many occasions, they want simple, inexpensive techniques suitable for the naive user, in the field. On other occasions, they may have a choice of laboratory methods available. The many implications of body composition states and the apparent simplicity of some of the measurements has led to the unquestioned use of techniques, particularly in field studies. However, the study of human body composition is complicated by methodological and biological issues, which are often conveniently put to one side in the heat of a survey or study and again when the results come to be interpreted.

This chapter considers the available body composition techniques and their applicability to the situations in which variability in human fertility is being investigated and to changing body composition in general. It considers the less known and forgotten measurement issues, rather than providing a comprehensive account of body composition methodology. The methods suitable for two types of use in fertility studies, field methods which have proven to be the most popular and laboratory methods used, for example, in studies of exercising women, are described.

Methods of measuring body composition

The techniques available for the study of human body composition are shown in Table 10.1. They are grouped according to *in vivo* and *in vitro* techniques, to regional or whole body techniques and techniques that can

167

Table 10.1. *Methods of measurement of body composition*

In vitro
  Anatomical dissection: muscle, skeleton, adipose tissue, viscera
  Chemical analysis: water, fat, protein, mineral, carbohydrate

In vivo
  Whole body
    Density
    Water
    Elements; K, Ca, N, C by *in vivo* neutron activation analysis (IVNAA)
  Regional
    Thicknesses
      Skinfolds, subcutaneous adipose tissue
      Ultra-sound (A-mode), computed tomography (CT)
      Dual energy radiography (DER), e.g. DEXA
      Magnetic resonance imaging (MRI)
    Areas
      Skinfolds plus circumferences
      Ultra-sound (B mode), CT, DER, MRI
    Volumes (and Masses)
      Areas plus lengths
  Estimations
    Regional
      skinfolds→density→whole body composition
      near infra-red interactance (NIRI)
    Whole body
      body mass index (BMI)→body composition
      bio-impedance analysis (BIA)
      total body electrical conductivity (TOBEC)

be used to estimate whole body composition. The most common whole body techniques are densitometry and hydrometry. The most common regional techniques are skinfold thicknesses and circumferences and the most common estimation procedures are based on skinfold thicknesses, and more recently, bio-impedance analysis (BIA). These procedures, and the assumptions on which they are based, have been reviewed extensively (Burkinshaw, 1985; Lukasi, 1987; Shephard, 1991; Elia, 1992; Lohman, 1992). The review by Jebb and Elia (1993) is a valuable short account of the practical considerations. Estimation procedures of body fat and fatness have been reviewed by this author (Norgan, 1991*a*; Norgan, 1995). Many of the other techniques are uncommon or have no place in fertility studies, particularly in the field. *In vitro* studies would have a place only in confirming the validity of new or existing *in vivo* techniques. Many techniques, such as *in vivo* neutron activation analysis (IVNAA), total body electrical conductivity (TOBEC) and computed tomography (CT) are limited to specialized clinical or medical physics laboratories.

Table 10.2. *Levels of measurement of body composition*

Level	Examples	Type
Mensuration	Dissection, chemical analysis	Direct
Transformation	Densitometry	Indirect
	Hydrometry	
Estimation	Anthropometry	Double indirect, calibrated
	Bio-impedance analysis (BIA)	against the above
	Near infra-red interactance (NIRI)	
Imaging	Computed tomography (CT)	Mensuration plus
	Dual energy radiography, e.g. DEXA	transformation
	Magnetic resonance imaging (MRI)	

Table 10.3. *Considerations in choosing a method*

Rationale
• dependent or independent variable

Precision and accuracy
• who is doing the measurement
• sample size
• changes or differences expected
• reference data

Apparatus
• cost
• availability
• technical difficulty
• maintenance and calibration

Acceptability by subjects
• ease and comfort
• modesty
• time

It is worth emphasizing that 'measurement' of body composition *in vivo* is always indirect and usually doubly indirect. Table 10.2 shows a description of the common techniques as mensuration, the *in vitro* techniques; transformation, where a physical property such as density is converted to composition with various biological and technical assumptions; and estimation where the method is calibrated against the above techniques with a further series of assumptions, particularly statistical. The newer imaging techniques involve mensuration and transformation but have less dependency on assumptions common in estimation techniques.

Some of the considerations in choosing a method are listed in Table 10.3. What usually determines the method chosen is the location: laboratory or field, developed or developing country; the number of individuals to be studied; or the apparent level of training required. These can be summarized as ease of use. Only more rarely and recently have the characteristics of measurements, the precision (reliability or repeatability), the accuracy (how close the result is to the true value) and the validity (how close what is being measured is to what is wanted) and the study requirements, i.e. the values or changes expected, been used to assess and chose the body composition method. These have been studied extensively for skinfolds and densitometry and to a lesser extent hydrometry but mainly in young adults. The precision and accuracy of a method depends on the setting, who is using it and under what conditions, laboratory or field.

Field techniques

In many studies of variability in human fertility, the subjects are women in rural areas of third world countries, often under nutritional or seasonal stress. Here, simple, inexpensive methods that are readily accepted by the population are most often required.

Body weight

The rationale for body weight measurements in this context is that they are a measure of overall size. Body weight is made up of lean body mass and adipose tissue, the latter being the most variable. Body weight is then a proxy for fat and fatness and changes in body weight proxies for changes in energy stores.

The measurement of weight can be simple, quick, precise and accurate and the apparatus portable and inexpensive. The simplicity of the measurement often leads to good technique being overlooked. As with all measurements, training, standardization (amount of clothing, voiding of bladder and bowels, time of day) and calibration of the apparatus remain of great importance (Weiner & Lourie, 1981; Lohman, Roche & Martorell, 1988; Blanchard, 1990). Spring-loaded 'bathroom' scales are not sufficiently accurate or robust in calibration, and have no place in these studies. Bias from any of these causes can contaminate the data. Random bias may not affect means but can make it difficult to show relationships between variables.

Unfortunately, body weight is a far from satisfactory measure of body composition or energy stores. Indeed, the major stimulus to the field of body composition was the unsatisfactory performance of body weight and weight: height indices in assessing fat stores and obesity. The story of extremely muscular American Football players being classified as over-weight and unfit for military service as a further stimulus is well known but apocryphal (Friedl, 1992). Not withstanding this, weight has proved to be a popular measurement, particularly for serial measurements in the field and small changes of less than 1 kg have been reported to affect ovarian function in seasonality studies (Panter-Brick, Lotstein & Ellison, 1993).

A second problem is that although the measurement of body weight is precise, the day-to-day variability complicates the interpretation of measurements and changes. For example, in a total of 2087 weight changes in young men the mean deviation (i.e. sign ignored), was 0.42 SD 0.23 kg. Thirty two percent were of 0.5 kg or more and 6% of 1 kg or more (Edholm, Adam & Best, 1974). Weight changes were correlated with energy expenditure ($r = -0.30$), energy intake (0.30) and fluid balance (0.35). There are less data for women and interpretation is complicated by variations during the menstrual cycle, but the mean changes and the percentages were less in women than in men, 0.28 kg, 13 and 2%, respectively (Robinson & Watson, 1965).

Variation in hydration seriously interferes with the interpretation of small changes in weight as measures of fat and energy stores. Garrow (1974) has shown convincingly that day-to-day variations in weight are associated with variations in the glycogen–water pool with an energy equivalent of 1 kcal/g. Day-to-day changes in body weight changes in the Tropics have been shown to be almost entirely due to water balance changes (Doré *et al.*, 1975). Changes in body water are also common during the menstrual cycle and have been measured (Watson & Robinson, 1965). On average, they are small but individual differences are marked. Changes up to 5 litres have been reported in women with and without reported symptoms of premenstrual tension (Andresch *et al.*, 1978). Thus, short-term changes in body weight are almost certainly changes in body water alone or a component with low energy content and should be interpreted cautiously. Even changes measured over long intervals can be affected by these day-to-day changes. Variability in hydration is a problem which affects all *in vivo* methods of body composition, as will be discussed.

Seasonal changes in body weight of up to 5 kg have been reported in different populations but the common experience is for means nearer 2 kg (Ferro-Luzzi & Branca, 1993). This is usually interpreted as changing energy stores, but few attempts have been made to control for level of hydration.

A further problem is that as weight is gained or lost, the composition of the tissue lost changes, in terms of fat and fat-free tissue, and hence the energy content changes (Grande, 1961). Also, Forbes (1987) has suggested that the composition of weight gain or loss and hence energy equivalence depends on the initial level, with a greater proportion of lean in the change of lean individuals. This has been confirmed with newer data (Prentice *et al.*, 1991). What all this shows is that weight or weight change, particularly small, short-term changes, are likely to be poor proxies for fat or energy store changes.

Other anthropometric indices

Body mass index (BMI)

The rationale of the body mass index and other weight:height indices is that weight will be a better indicator of fat and energy stores if an allowance is made for size by, in this case, dividing by height squared. BMI is currently receiving much attention and has been adopted by the Food and Agricultural Organization of the United Nations as a measure of adult chronic energy deficiency (Shetty & James, 1994). More and more data are appearing on its relations to functional indices, such as mortality and morbidity, and work output. BMI is superior to other W:H indices, in terms of its relationship with body composition, and good reference data exists. However, its interpretation is not straightforward. It seems to be a better measure of fat mass than fatness which should be an advantage. It is as much a measure of leanness as of fatness and of size as much as of composition. Its relationship with fatness and fat content differ with age and in various ethnic groups (Norgan, 1990, 1994*a*) and, for some, its interpretation may be affected by the body shape as evinced by the SH/S ratio (Norgan, 1994*b*). A charitable view of BMI is that it seems to have suffered from design drift and while it does not meet the original specifications of a measure of body fatness, it is still a valuable measure. Alternatively, the process can be regarded as one of interactive design, that is changing its interpretation with experience. Nowadays, the description of subjects should include the BMI to indicate their approximate energy nutritional status and to a compare the results of various studies but care is required when interpreting BMI as a measure of energy stores. Changes in BMI follow changes in weight in adults.

The use of equations with height and weight as predictor variables to estimate minimum levels of fatness for menarche and menstruation in early studies has not been substantiated (Reeves, 1979; Katch & Spiak, 1984).

Mid upper arm circumference (MUAC)

MUAC is an established measurement in the assessment of the nutritional status of children and of hospital patients. Good reference data exists. Efforts to improve the value of the data by transformation to arm muscle area (AMA) and arm fat area (AFA) have not been successful. This is usually explained by the technical difficulties of skinfold measurements and the assumptions made about the geometry of the upper arm. MUAC is highly correlated with BMI in adults. This is surprising at first but like BMI it is foremost a combination of vectors of size and then of adipose and lean tissues. James *et al.* (1995) have shown how in many cases MUAC alone can be used as a measure of adult chronic energy deficiency (CED) without the need for height and weight apparatus and measurement and how it can be used to refine the categories of CED based on BMI.

Some other body circumferences are used as a simple measures of fat distribution, the most well-known being the waist–hip circumference ratio (WHR). A recent introduction is the conicity index of Valdez *et al.* (1993) for assessing abdominal fatness. The index is given by abdominal circumference $/ (0.109 \sqrt{\text{weight/height}})$. The model implies that abdominal fatness is represented by a progression from a perfect cylinder, whose circumference is generated by the height and weight, to a double cone, and hence it has a finite range. It could be regarded as following Cezanne's dictum that we see nature as cones, shapes and spheres! The index is more strongly associated with cardio-vascular risk factors than WHR in men and has the advantages that it is not necessary to measure hip circumference. It adjusts waist circumference for height and weight, allowing comparison with other individuals and groups.

Skinfold thickness

Skinfold thicknesses have many of the characteristics of a good field technique. They are simple and quick to take, the apparatus is cheap and portable and reference data is plentiful. However, skinfolds are considered to be of low precision, below those of other field and bedside methods such as bio-impedance analysis and near infra-red interactance but, considering the variation in skinfold thicknesses between individuals, the precision that can be obtained after good training and with continuous quality control is acceptable in most cases (Oppliger & Spray, 1987). Although there is much to be said for interpreting skinfold thicknesses by comparisons with reference data, their use to estimate whole body fatness has proved to be almost universally irresistible.

Most of the estimation equations have been drawn up on Europeans and have proved to be population specific. Specificity may be methodological or biological in origin (Norgan & Ferro-Luzzi, 1985). Whereas the former, arising from technical or measurement factors and statistical considerations, can be reduced or controlled, the former are difficult to diminish without great effort. The Durnin and Womersley (1974) estimation equations have proved to be most popular and robust. They are sex and age specific and include most combinations of the biceps, triceps, subscapular and suprailiac skinfolds as log sums to linearize their relationship with body density. Deriving age specific equations, rather than incorporating age into the equations, causes step changes in body composition for a given sum of skinfolds at different ages. This is apparent in Fig. 10.1 which shows the percentage fat calculated from the age-specific equations for a given total skinfold of 68 mm. The Figure also shows that based on the data in Durnin and Womersley (1974) in European women estimates of percentage fat from 2, 3 or 4 skinfolds are similar. (The exception is for sum of three skinfolds for 16–68 year olds. This suggests a mistake in the original formula.) However, using data of Papua New Guinean women from Ulijaszek *et al.* (1989), there are marked differences when using different skinfold combinations owing to differences in subcutaneous fat distribution.

Patterson (1992) used validity generalization, a meta-analysis model, to ascertain the generalizability of skinfolds as measures of body density and hence body fatness. No single site was generalizable to all subjects but the log sum of the four skinfolds used in the Durnin and Womersley (1974) equations was. The subscapular site was generalizable to women and the suprailiac and midaxillary sites to men.

Oppliger *et al.* (1987) concluded that skinfolds were highly reliable and there were no technical reasons arising from skinfolds why estimation equations should not give accurate estimates. They found by generalized analysis of variance that the measurement error of skinfolds in two common estimation equations in their hands contributed only 5–10% of the body density prediction error in men (Oppliger & Spray, 1987). Whatever their faults, skinfolds stand up to scrutiny compared with other field methods (Pullicino *et al.*, 1990; Pierson *et al.*, 1991; McNeil *et al.*, 1991). They are as good as three component models, DEXA and the other field methods as estimates of mean percentage fat of groups from four component models and the best of the field techniques for individual estimates (Fuller *et al.*, 1992).

Special consideration is required for several groups. Cumming and Rebar (1984) compared five estimation equations in three groups of six

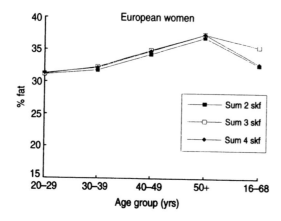

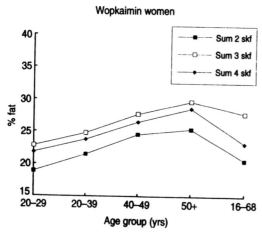

Fig. 10.1. The effects of age on the interpretation of a constant sum of skinfold thickness according to the age-specific estimation equations of Durnin and Womersley (1974) and the effects of differences in subcutaneous fat distribution, as shown by Wopkaimin women of Papua New Guinea (Ulijaszek *et al.*, 1989), on the estimates.

normal control women, six normally menstruating and six amenorrheic runners. Significant differences between the groups were found for some but not all the methods demonstrating the importance of choice of method on outcome. Validated equations for adolescent and women athletes have been produced (Thorland *et al.*, 1984; Sinning & Wilson, 1984). Changes in fat distribution during pregnancy may invalidate skinfold estimation techniques (Norgan, 1992) but not apparently during lactation (Butte *et al.*, 1985; Wong *et al.*, 1989). Any estimation technique, skinfolds or others, ought to be validated on a sample of the group under investigation before general application.

There has been little consideration of the applicability of skinfold estimation equations to non-Europeans. In a recent review, Norgan (1995) concluded that the estimation equations of Durnin and Womersley (1974) appeared to be the most appropriate for non-Europeans in that in half the studies in the literature the mean estimated and measured densities were not significantly different. Most of the studies were on young men. The other common regression equations such as those of Jackson, Pollock and colleagues (Jackson & Pollock, 1978; Jackson, Pollock & Ward, 1980) had not been as widely tested in the same way but where they had, they performed less well than those of Durnin and Womersley (1974).

Bio-impedance analysis (BIA) and near infra-red interactance (NIRI)

BIA

This relatively new technique has gained widespread acceptance, judging from the numbers of papers published on it or using it, to the extent that a bioimpedance 'craze' has been described (Elia, 1993). The most recent review at the time of writing lists over 120 papers (Heitmann, 1994). This has less to do with the technique being significantly better than other field techniques but more to do with its ease of use and acceptability. It has been reviewed frequently and critically (Baumgartner, Chumlea & Roche, 1990; Heitmann, 1994). The principle of the method is that the electrical impedance of the body to a weak alternating electrical current (800 μA, 50 kHz) applied through source electrodes at the wrist and received at the ankle is related to the total body water (TBW). Total body water is related to a fat-free mass and hence indirectly to fat mass. The key assumptions are that body geometry is a simple cylinder, that body temperature is constant and electrolytes and water are uniformly distributed. Impedance is proportional to the length of the conductor, a function of height, and, indirectly, to the cross-sectional area of the body, rather than to the mass such that an arm may make up 5% of the body weight but up to 45% of the impedance. Individual and ethnic differences in the relative lengths of the limbs may compromise BIA as they do BMI (Norgan,1994b). The frequency of the current determines whether it penetrates extracellular water alone or whether it crosses cell membranes into intracellular water too. Multi-frequency methods are being developed to measure both extracellular and total body water.

Measurements of impedance and reactance are related to TBW by estimation equations which also include anthropometric variables to reflect the geometry of the body. Here, and with the use of NIRI, there is much debate about the relative contributions of height and weight compared with the physical property in question to body composition with several groups finding little extra being added to the anthropometry (Diaz *et al.*, 1989, 1991; Deurenberg *et al.*, 1991; Eckerson, Housh & Johnson, 1992) and others that height^2/resistance is the best single predictor of TBW (Heitmann, 1990; Kushner *et al.*, 1992). There are said to be 30 different manufacturers of BIA machines, and it has been a common experience that the manufacturer's estimation equations are not widely applicable, usually underestimating fat mass in the obese and overestimating it in the lean (Heitmann, 1994). Different makes of machine (Graves *et al.*, 1989) and even the same types of machine from the same manufacturer have been found to perform differently (Deurenberg, van der Kooy & Leenen, 1989*a*). Experience suggests that a single equation to convert impedance to body composition will not be appropriate for all subjects, and that investigators should assess the validity of existing equations. For some groups it may be necessary to develop new estimation equations but it is felt advisable to avoid this wherever possible as it can be a major undertaking and if performed unsatisfactorily as is often the case the literature becomes confused and almost uninterpretable.

As a measure of body composition, BIA is affected by hydration. It may be measuring TBW accurately but the relationship of TBW to FFM is altered. It is advisable to standardize measurement conditions by measuring after an overnight fast, and 24 hours without alcohol or exercise (Deurenberg, Weststrate & Paymes, 1988; Heitmann, 1994). This is rather restrictive for a good field technique, particularly for rural third world women. The method has been considered reliable for measurements over the menstrual cycle, albeit confounded by small weight changes related to hydration status (Gleichauf & Roe, 1989).

BIA performs less well at the ends of the range of body composition (Malina, 1987), like most estimation techniques and has been shown to be fatness dependent (Segal *et al.*, 1988). Fatness specific equations reduce the standard error of estimate (SEE) to 2-3 kg LBM. Age and sex specific cross-validated equations have been produced but these have been considered little better than simple anthropometric techniques alone by some (Deurenberg *et al.*, 1991) but not others (Heitmann, 1994).

Are the BIA equations applicable to non-European women? The current methods for impedance analysis and the transformation of the data have been shown repeatedly to give different means to those from densitometry and hydrometry in groups other than Europeans (Norgan, 1995). The method has been shown to be accurate in Pima Indians, an obese population, using a locally developed equation (Rising *et al.*, 1991). Some success has been achieved with the method in pre-menopausal Chinese women (de Waart, Li & Deurenberg, 1993).

BIA is popular because it is simple to use, although the importance of electrode positioning particularly on the wrist is often overlooked. It is quick and involves minimum interference with or undressing by the subject and the instruments produce results without the need for further transformation or calculation. BIA has been adopted in the third US National Health and Nutrition Examination Survey (NHANESIII) for fatness measures in adults in conjunction with four skinfold thicknesses and four circumferences. BIA's better reproducibility than skinfold thicknesses was a point in its favour in this study of 45 000 individuals by many different observers. BIA has proved reliable in inter-laboratory comparisons of the technique (Segal *et al.*, 1988; Deurenberg *et al.*, 1994).

NIRI

The principle of this technique is that when tissues are irradiated with infra-red radiation the optical density of the reflected radiation depends on the absorption characteristics of the underlying tissues, adipose and lean varying in this respect. The common commercial instrument, Futurex 5000 (Self Care Products Ltd, Amersham, Bucks) uses infra-red radiation of two wavelengths 940 and 950 nm. As with BIA, the physical properties are related to the biological by estimation equations. In the case of the Futurex 5000, height, weight, sex and level of physical activity are included. Measurement is made at a single site, the biceps, but measurements at additional sites add little to the estimation.

Most workers have found that NIRI has little advantage over skinfold thicknesses and other anthropometric approaches (Israel *et al.*, 1989; Elia, Parkinson & Diaz, 1990; Hortobagyi *et al.*, 1992; McLean & Skinner, 1992; Fuller *et al.*, 1992). In our experience with middle-aged men it performed poorly, with a bias (mean difference) of −10% fat and with increasing underestimation with higher percentage fat (Brooke-Wavell, 1992) but new software has improved this by reducing the bias but not below the level of skinfolds (Brooke-Wavell *et al.*, 1995). This highlights the point that developments in hardware and software can improve performance such that a review of the literature may not reflect the current capabilities of an analyser. A Futurex 6000 which measures at six wavelengths is now on the market.

Laboratory-based techniques

The laboratory-based techniques of densitometry and hydrometry have been in use for many years and are not considered in detail here. Several recent reviews have considered the methods and their limitations as two component models of body composition, usually fat and fat-free or adipose tissue and lean body mass (Lukaski, 1987; Coward, Parkinson & Murgatroyd, 1988; Elia, 1992; Lohman, 1992). The assumptions of constant composition and proportions of the components of the lean tissue does not reflect accurately the true situation. The problems of varying hydration, that may be particularly important in field studies in the Tropics, have been referred to earlier. The fertility of women athletes with their high energy turnovers and low energy stores has been the subject of much investigation but competitive middle and long distance women runners are usually light and lean, differing from the normal population in having lower fat masses, lighter bones and often lean masses too. These too can alter the composition of the fat-free mass such that densitometry may underestimate and hydrometry may overestimate fatness and energy stores. Whereas the effects of variations in bone mineral content on estimating percentage fat from densitometry for most young women is modest, for those athletes with extremely high (body builders) or extremely low (amenorrheic runners) bone mineral content may require adjustments in the calculation of body composition (Bunt *et al.*, 1990). There is no doubt that multicomponent models of body composition will better describe the true picture (Coward *et al.*, 1988). The size of the gain in information may, in many cases, be small in relation to the 'costs' off the extra measurements.

Combining densitometry and hydrometry has enabled the body to be considered as a three component system but a major advance has been accessible methods for total body mineral. These overcome many of the problems that arise from the variability in composition of the different compartments of the body within and between individuals and during the lifespan. However, few laboratories have open access to all the techniques necessary for multicomponent methods and the two-component methods will continue to play a role in the description of body composition in most studies.

The major role of the multicomponent methods will be in validating newer techniques and estimation equations and the newer imaging techniques will provide data on regional composition and organ size. It must be said however, that each technique has required some validation process and often the newer multicomponent techniques themselves have been validated by two component methods.

Densitometry

If the body is regarded as two compartments and the density of the body (D_b) and the densities of its two constituents are known the proportions of the constituents can be calculated. The two constituents are usually taken to be fat (F) and fat-free mass (FFM). Hence,

$$1/D_b = (f/D_f) + (ffm/Dffm)$$

where f is the proportion of fat and ffm the proportion of fat-free mass. Their densities have been estimated from animal studies, the individual constituents and, in a few cases from cadaver analyses (Siri, 1961). The complication in the measurement of body density is how to measure the volume of an irregularly shaped body. This is usually achieved using Archimedes' principle of buoyancy, the loss in weight of a submerged body is equal to the weight of water displaced by the body, from which the volume of the body can be calculated. In humans, an allowance must be made for the air in the lungs. Full experimental details are given in Norgan (1991b) but the precise details vary from system to system (Jebb & Elia, 1993). The maximum range of density values is only some 10% so accuracy, particularly in the measurement of residual lung volume, is important. The measurement of body density is highly repeatable, (Bartoli *et al.*, 1993) such that precision (SD of differences) is equivalent to 0.7 kg fat (Coward *et al.*, 1988). In this respect, it out-performs all the other techniques (Lukaski, 1987).

Lohman (1981) suggested that the largest contribution to the error of estimating % fat from density comes from the variability of body water. Variations in weight that are due to water are liable to misinterpretation by density measurements. A 1% change in body water appears to be a 0.7% units change in % fat (Martin & Drinkwater, 1991). Girandola, Wiswell and Romero (1977) found that density measurements following the ingestion of 1.8 litres of fluid suggested a gain of fat of more than 1% as fat and a kg weight loss from sweating a fall more than 0.7%.

The variations in body weight over the menstrual cycle are often ascribed to body water changes. Svoboda and Query (1984) measured body density three times a week and on the day of maximum and minimum weight in 26 women over the menstrual cycle. The mean peak difference in body fat was 0.3%, a non-significant change, but in some individuals this was much more. The within subjects peak difference was 1.7% fat. However, Bunt, Lohman and Bioleau (1989) concluded that changes in total body water could explain only part of the significantly different densities in women who experienced perceptible changes in body weight over the menstrual cycle.

The remaining differences may have been due to changes in fat and protein content or methodological errors. The usual equations for converting body density or TBW to percentage fat are not appropriate during pregnancy due to the changing relative proportions of the constituents although validated equations have been developed (van Raaij *et al.*, 1988).

Density has been measured in several studies of exercise and menstrual disturbances. Calabrese *et al.* (1983) investigated 34 ballet dancers, 17 with menstrual abnormalities. There was no difference in percentage fat in those with and without menstrual abnormalities. The now discarded Frisch & McArthur (1974) method of assessing percentage fat from height and weight gave significantly different values from those of densitometry. Sanborn, Albrecht and Wagner (1987) found no association between athletic amenorrhoea and body fat and Boyden *et al.* (1982) found menstrual dysfunction with exercise but without weight loss or extreme leanness, using densitometry.

For one reason or another, densitometry is regarded as the best single laboratory technique. Not all subjects may be able to complete the immersion procedures, however, and in this respect hydrometry has an advantage.

Hydrometry

The rationale of hydrometry as a means of measuring body composition is based on the premise that lean tissue contains a constant known amount of water and if the total body water can be calculated so too can the fat-free mass. Total body water is usually measured by the dilution of labelled water, such as the stable, non-radioactive isotope of hydrogen, deuterium, and by sampling body fluids such as sweat, saliva or urine. The analytic standard error of about 1 l is equivalent to 1.4 kg FFM, and, by difference, 1.4 kg fat, and in this respect hydrometry does not match that of densitometry. Bartoli *et al.* (1993) have determined the variability in four weekly measurements of TBW by deuterium dilution and found a CV of 4% or 1.8 l, similar to underwater weighing. Generalizability ANOVA suggested that 25–50% of the variability in TBW was due to technical factors. Thus, changes of TBW of less than 0.5–0.9 l would be within the error of the method. Most other studies of repeatability have used only two measurements over varying intervals but most find body weight and water changes in the same direction and of the same order of magnitude. Of the other isotopic methods, ^3H_2O is radioactive and repeated testing has to be limited and $H_2^{18}O$ is very expensive. However, they are likely to be similar in repeatability. A drawback of the method is the 3–5 h equilibration period required after administration of the diluent. This may be inconvenient.

Dual energy radiography (DER)

Dual energy radiography (DER) is a new technique for whole body and regional body composition that is being used increasingly in studies in medicine and biology. The method is well established for bone mineral but still being developed as a reliable method for soft tissue (Elia, 1992; Roubenoff *et al.*, 1993). DER began with dual photon absorptiometry (DPA) for measurement of bone mineral but this had a number of problems. Newer instruments have an X-ray source (as opposed to a ^{153}Gd source) with two main energy beams. The principle is that soft tissue and bone attenuate photons (or X-rays in the case of dual energy X-ray absorptiometry, DEXA) to differing degrees. Photons of two energies, some 40 and 80 keV, allow separation of bone mineral from soft tissue and fat from lean. The proportion of adipose and lean in each pixel can be calculated from the attenuation of pure fat and bone free lean plus the ratio of soft tissue attenuations at the two beam energies. The original use of the soft tissue data was to correct the attenuations and improve the precision of bone measurements but DEXA has been developed as a technique for adipose and non-bone lean tissue.

For bone, precisions (CV of repeated measurements) of 2–3% and accuracies of 1% have been reported. For percentage fat the precision was 1% (Jebb & Elia, 1993) but the accuracy of the method for fat is not well established. The usual approach of comparison with an established method, densitometry, has shown it to perform adequately (Van Loan & Mayclin, 1992; Johansson *et al.*, 1993). Another approach has been to sum the separate compartments and to compare this with body weight. The sum of the compartments is estimated accurately but the individual compartments may not be.

DEXA is not free from the assumptions about hydration (Roubenhoff *et al.*, 1993) or the proportions of protein to water. The size of the effect is not known but Going *et al.* (1993) report that DEXA is unaffected by changes of 1–2 litres in hydration. In this respect, DEXA would be better than densitometry. However, DEXA cannot distinguish clearly between soft tissue and bone in the thorax, because measurements are affected by the anterior-posterior thickness. Also, in the arm where the proportion of bone is high, the error for soft tissue is high. Pixels with small amounts of bone appear as 'very lean', overestimating leanness, which is also a problem with pixels of large size, as used in the whole body scans. Difficulties are experienced accommodating large individuals in the scanner, particularly the obese. Software upgrades have been frequent, requiring reanalysis of data and it is not known how machines from different manufacturers compare.

DEXA has been described as a boon for bone and soft tissue measurement but it is too soon to claim it as a gold standard method for soft tissue (Roubenoff *et al.*, 1993). Its advantages are that it is quick, 10–15 min for total body scan and that the radiation dose is low, less than 5 mrem. The analysers are not restricted to radiology departments.

DEXA can give the distribution of adipose tissue within an area, subcutaneous versus visceral adipose, for example, and between areas such as the limbs and trunk. It is becoming apparent that some sites may be more mobilizable than others, particularly in pregnancy and lactation, and that some sites of fat deposition pose a greater risk to health than others, so imaging techniques, particularly for visceral fat, are likely to be prominent in the near future, but for reasons unrelated to fertility.

Computed tomography (CT)

This technique involves radiation doses at higher levels than DEXA and consequently does not have the same widespread use. In essence, X-ray sources and detectors are located on a circular gantry into which the body is placed. Complex reconstructive software is used to generate cross-sectional areas and, with several scans at other sites, to build up volumes and masses. Issues of slice width and inter-slice distance are important as are the assumptions in changing areas to masses. Subject movement must be minimal during the 5–10 s it takes for each scan. The method has been validated using phantoms or by comparison with standard methods such as densitometry and has been used to establish estimation equations for visceral as well as total adipose tissue volumes (Kvist *et al.*, 1988).

Magnetic resonance imaging (MRI)

Some elements such as hydrogen, phosphorus, carbon and sodium have nuclei with odd numbers of protons or neutrons such that when surrounded by a magnetic field they align themselves relative to the field. When the field is withdrawn, the energy of the nuclei is dissipated and can be recorded and processed to give cross-sectional areas like CT scans. No ionizing radiation is produced so the method is acceptable for repeated measurements in women and children. One problem with MRI, however, is the slow scan time, 10–15 minutes per scan, during which time the subject must remain still. Mean differences for percentage fat of 1% and for visceral fat of 5% have been reported (Ross *et al.*, 1993; Sohlstrom, Wahlund & Forsum, 1993) and the coefficient of variation some 5% and 10–15% respectively (van der Kooy & Seidell, 1993). MRI has been shown to be accurate in assessing fat and adipose tissue of pigs (Fowler *et al.*, 1992). Frisch *et al.* (1993) have used MRI to examine subcutaneous and internal fat between the 5th thoracic and femoral regions in athletes and controls and added to the information on fatness and menstruation without uncovering any major new insights. The ratio of subcutaneous to internal fat was 80:20 in all groups. Athletes with menstrual disorders had significantly less subcutaneous and internal fat than controls and 3-oestradiol 2-hydroxylation was inversely related to percentage fat and subcutaneous fat in each section.

These newer methods have been reviewed by Fuller *et al.* (1990) and a good practical guide to imaging techniques is given by van der Kooy and Seidell (1993).

Measuring changes in body composition and the effects of changes on the determinations

In many cases, investigators are particularly interested in changes in body composition and body energy content but the ability of most body composition techniques to measure changes in body composition is reduced. The problem is that the changes themselves may reduce the accuracy of a method by causing violations in the assumptions on which it is based. For example, decreases in mineral or muscle tend to result in underestimation of fatness and decreases in water tend to overestimate. Much of the work on measuring changes has been performed on the obese undergoing reduction and such findings may not be applicable in this context.

Of the field methods, Almond *et al.* (1984) suggested that estimates of changing body composition from skinfolds during intravenous feeding were as good as high-technology approaches, such as IVNAA. In recent years, most attention has been given to BIA for the measurement of changes. Deurenberg *et al.* (1989*b*) found that, after short-term weight loss of 1 kg, BIA overestimated FFM loss by 1 kg in healthy volunteers. Jebb and Elia (1991) studied eight patients undergoing renal dialysis and found BIA overestimated fluid removal by 86–100%. The relationships between BIA changes and fluid changes were very different in each individual. Webber & Macdonald (1992) concluded that BIA could measure the group changes in body composition with a 72 h fast provided hydration was maintained. Kushner *et al.* (1990) championed the cause of BIA against skinfolds in the measurement of changes but others have regarded it as unreliable (Deurenberg *et al.*, 1989*a* and *c*) or no better than weight (Mazess, 1991). By way of a summary, Forbes, Simon and Amatruda (1992) combined the results from 7 studies on 98 subjects and concluded that weight change was a more reliable predictor of LBM change than impedance change.

Jebb *et al.* (1993) compared a number of different methods to test their ability to measure changes in body composition in six under- or overfed subjects compared with fat balance from 12-day continuous whole body calorimetry. They concluded that a three-component method involving measurements of density and TBW was best, based on the least bias (mean difference in the methods) and greatest precision (lowest SD of the differences). As the greatest precision of any method was 0.77 kg, they suggest that the smallest change in fat that can be measured is 1.54 kg. It is noticeable that BMI and skinfolds have better biases if not precisions, i.e. they are better for estimating mean changes but less successful at measuring individual changes. Earlier, Garrow *et al.* (1979) found densitometry to be the best of the two component methods in the reducing obese.

Are re-interpretations of the literature necessary?

It is intuitively acceptable that body composition, in particular the amount of body fat and hence the level of energy stores, should be linked to successful reproduction or fertility. Reproduction is crucial to the survival of the species but expensive in energetic terms for mammals, particularly primates. There is now much evidence of an association between level of nutrition, particularly energy nutrition as evinced by the body fatness, and the onset and maintenance of a reproductive capability. This obtains for the rate of maturation, age at menarche, maintained ovulation and lactational amenorrhoea. Evidence comes from periods of war and famine, secular improvements in health and nutrition, and behaviours such as participation in distance running, ballet dancing and eating disorders. Most of this evidence is correlational, i.e. associative, but the idea of a causal relationship has its supporters, and putative mechanisms have been described for the effects.

In view of the requirements and assumptions of the methods described, do current ideas about a causal effect of body composition on fertility need to be re-assessed or re-interpreted? Certainly, there is a case for more parsimony in interpreting evidence in favour of the existence or position of a minimum level of energy stores for all individuals and populations below which fertility is invariably compromised or above which fertility is unaffected by biological considerations. Firstly, for a group of individuals with the same energy stores the methods would produce a range of compositions due to biological variability. Secondly, estimates at the ends of the range tend to be more biased than those in the middle and thirdly changing composition often invalidates the assumptions of the methods or estimates. The accuracy of body composition methods needs to be borne in mind. Even under the best circumstances, the errors are more than 0.5 kg fat or some 4500 kcal and for measuring changes are likely to be more than 0.75 kg (Elia, 1992). The errors of estimation techniques used in the field will be greater because most have been established against two compartment models and methods and because of their greater inter-observer variability.

Summary and recommendations

The preceding sections can be summarized as follows:

(i)   Body weight is not a reliable estimator of body composition. Body weight changes, particularly short-term changes of 0.5–1 kg, may reflect changes in hydration, not energy stores.

(ii)   If estimations have to be made, the equation used should be validated on the population under study or a widely validated equation chosen, such as one of those of Durnin and Womersley (1974) which have been shown to have the lowest bias and highest precision.

(iii)   In spite of the many studies on BIA, further validation is required for many groups.

(iv)   Densitometry and hydrometry can be affected by variations in hydration.

(v)   Multicomponent methods, involving densitometry, hydrometry and DEXA to measure fat, water and mineral, promise to perform best but are not widely available.

(vi)   Few of the methods have been validated on groups other than Europeans and those of European descent.

Recommendations of three levels of measurement, basic anthropometry, extended anthropometry and desirable measurements to meet the needs of most studies are given in Table 10.4.

Table 10.4. *Suggested measurements of anthropometry and body composition in studies of the biological factors influencing variability in human fertility*

Types of measurement	Field measurements		Laboratory Desirable
	Basic	Recommended	
Mass	Body weight	Basic plus	Recommended plus
Linear	Stature		Sitting height,
Skinfolds	Triceps and subscapular	Supra-iliac, thigh, biceps, abdomen	Calf, trunk, mid-thigh
Circumferences	Upper arm	Waist, hip, upper thigh	Mid-thigh, calf
Body composition		Estimates (Durnin & Womersley, 1974, BIA	Density Body water DEXA

REFERENCES

Almond, D. J., King, R. F. G. J., Burkinshaw, L., Oxby, C. B. & McMahon, M. J. (1984). Measurement of short-term changes in the fat content of the body. A comparison of three methods in patients receiving intravenous nutrition. *British Journal of Nutrition*, **52**, 215–25.

Andersch, B., Hahn, L., Andersson, M. & Isaksson, B. (1978). Body water and weight in patients with premenstrual tension. *British Journal of Obstetrics and Gynaecology*, **85**, 546–50.

Bartoli, W. P., Davis, J. M., Pate, R. R., Ward, D. S. & Watson, P. D. (1993). Weekly variability in total body water using (^{2}H$_2$O) dilution in college-age males. *Medicine and Science in Sports and Exercise*, **25**, 1422–8.

Baumgartner, R. N., Chumlea, W. C. & Roche, A. F. (1990). Bioelectric impedance for body composition. *Exercise and Sports Science Review*, **18**, 193–224.

Blanchard, J. (1990). The importance of accurate and standardised height and weight measurements in body composition analysis. *European Journal of Clinical Nutrition*, **44**, 337–8.

Boyden, T. W., Parmenter, R. W., Grosso, D., Stanford, P., Rotkis, T. & Wilmore, J. H. (1982). Prolactin responses, menstrual cycles, and body composition of women runners. *Journal of Clinical Endocrinology and Metabolism*, **54**, 711–14.

Brooke-Wavell, K. (1992). Human body composition: measurement and relationship with exercise, dietary intake and cardio-vascular risk factors. Unpublished PhD Dissertation, Loughborough University of Technology, Loughborough, UK.

Brooke-Wavell, K., Jones, P. R. M., Norgan, N. G. & Hardman, A. E. (1995). Evaluation of near infra-red interactance for assessment of subcutaneous and total body fat. *European Journal of Clinical Nutrition*, **49**, 57–65.

Bunt, J. C., Goin, S. B., Lohman, T. G., Heinrich, C. H., Perry, C. D. & Pamenter,

R. W. (1990). Variation in bone mineral content and estimated body fat in young adult females. *Medicine and Science in Sports and Exercise*, **22**, 564–9.

Bunt, J. C., Lohman, T. G. & Boileau, R. A. (1989). Impact of total body water fluctuations on estimation of body fat from body density. *Medicine and Science in Sports and Exercise*, **21**, 96–100.

Burkinshaw, L. (1985). Measurement of human body composition *in vivo*. *Progress in Medical Radiation and Physics*, **2**, 113–37.

Butte, N. F., Wills, C., O'Brian Smith, E. & Garza, C. (1985). Prediction of body density from skinfold measurements in lactating women. *British Journal of Nutrition*, **53**, 485–9.

Calabrese, L. H., Kirkendall, D. T., Floyd, M., Rapoport, S., Williams, G. W., Weiker, G. G. & Bergfield, J. A. (1983). Menstrual abnormalities, nutritional patterns and body composition in female classical ballet dancers. *Physician and Sports Medicine*, **11**, 86–98.

Coward, W. A., Parkinson, S. A. & Murgatroyd, P. R. (1988). Body composition measurements for nutrition research. *Nutrition Research Reviews*, **1**, 115–24.

Cumming, D. C. & Rebar, R. W. (1984). Lack of consistency in the direct methods of estimating percent body fat. *Fertility and Sterility*, **41**, 739–42.

Deurenberg, P., van der Kooy, K. & Leenen, R. (1989a). Differences in body impedance when measured with different instruments. *European Journal of Clinical Nutrition*, **43**, 885–6.

Deurenberg, P., van der Kooy, K., Leenen, R., Weststrate, J. A. & Seidell, J. C. (1991). Sex and age specific prediction formulas for estimating body composition from bioelectrical impedance: a cross-validation study. *European Journal of Clinical Nutrition*, **15**, 17–25.

Deurenberg, P., Westerterp, K. R. & Velthuis-Te Wieril, E. J. M. (1994). Between-laboratory comparison of densitometry and bio-electrical impedance measurements. *British Journal of Nutrition*, **71**, 309–16.

Deurenberg, P., Weststrate, J. A. & Hautvast, J. G. A. J. (1989b). Changes in fat-free mass during weight loss measured by bioelectrical impedance and by densitometry. *American Journal of Clinical Nutrition*, **49**, 33–6.

Deurenberg, P., Weststrate, J. A. & Paymes, I. (1988). Factors affecting bioimpedance measurements in humans. *European Journal of Clinical Nutrition*, **42**, 1017–22.

Deurenberg, P., Weststrate, J. A. & van der Kooy, K. (1989c). Body composition changes assessed by bioelectrical impedance measurements. *American Journal of Clinical Nutrition*, **49**, 401–3.

De Waart, F. G., Li, R. & Deurenberg, P. (1993). Comparison of body composition assessments by bioelectrical impedance and by anthropometry in premenopausal Chinese women. *British Journal of Nutrition*, **69**, 657–64.

Diaz, E., Gonzalez-Cossio, T., Rivers, J., Immink, M. D. C. & Dario, R. D. (1991). Body composition estimates using different measurement techniques in a sample of Highland subsistence farmers in Guatemala. *American Journal of Human Biology*, **3**, 525–30.

Diaz, E. O., Villar, J., Immink, M. & Gonzales, T. (1989). Bioimpedance or anthropometry. *European Journal of Clinical Nutrition*, **43**, 129–37.

Doré, C., Weiner, J. S., Wheeler, E. F. & El-Neil, H. (1975). Water balance and body weight: studies in a tropical climate. *Annals of Human Biology*, **2**, 25–33.

Durnin, J. V. G. A. & Womersley, J. (1974). Body fat assessed from total body density and its estimation from skinfold thickness: measurements on 481 men and women aged from 16 to 72 years. *British Journal of Nutrition*, **32**, 77–97.

Eckerson, J. M., Housh, T. J. & Johnson, G. O. (1992). Validity of bioelectrical impedance equations for estimating fat-free weight in lean males. *Medicine and Science in Sports and Exercise*, **24**, 1298–302.

Edholm, O. G., Adam, J. M. & Best, T. W. (1974). Day-to-day weight changes in young men. *Annals of Human Biology*, **1**, 3–12.

Elia, M. (1992). Body composition analysis: an evaluation of 2 component models, multicomponent models and bedside techniques. *Clinical Nutrition*, **11**, 114–27.

Elia, M. (1993). The bioimpedance 'craze'. *European Journal of Clinical Nutrition*, **47**, 825–7.

Elia, M., Parkinson, S. A. & Diaz, E. O. (1990). Evaluation of near-infra red reactance as a method for predicting body composition. *European Journal of Clinical Nutrition*, **44**, 113–21.

Ferro-Luzzi, A. & Branca, F. (1993). Nutritional seasonality: the dimensions of the problem. In Seasonality and Human Ecology, ed. S. J. Ulijaszek and S. S. Strickland, pp. 149–65. Cambridge: Cambridge University Press.

Forbes, G. B. (1987). Lean body mass – body fat interrelationships in humans. *Nutrition Reviews*, **45**, 225–31.

Forbes, G. B., Simon, W. & Amatruda, J. M. (1992). Is bioimpedance a good predictor of body-composition change? *American Journal of Clinical Nutrition*, **56**, 4–6.

Fowler, P. A., Fuller, M. F., Glasbet, C. A., Cameron, G. G. & Foster, M. A. (1992). Validation of the *in vivo* measurement of adipose tissue by magnetic resonance imaging of lean and obese pigs. *American Journal of Clinical Nutrition*, **56**, 7–13.

Friedl, K. E. (1992). Body composition and military performance: origins of army standards. In *Body Composition and Physical Performance*, ed. B. M. Marriott and J. Grumstrup-Scott, pp. 31–55. Washington, DC: National Academy Press.

Frisch, R. E. & McArthur, J. W. (1974). Menstrual cycles: fatness as a determinant of minimum weight for height necessary for their maintenance or onset. *Science*, **185**, 949–51.

Frisch, R. E., Snow, R. C., Johnson, L. A., Garerd, B., Barbieri, R. & Rosen, B. (1993). Magnetic resonance imaging of overall and regional body fat, estrogen metabolism, and ovulation of athletes compared to controls. *Journal of Clinical Endocrinology and Metabolism*, **77**, 471–7.

Fuller, M. F., Fowler, P. A., McNeil, G. & Foster, M. A. (1990). Body composition: the precision and accuracy of new methods and their suitability for longitudinal studies. *Proceedings of the Nutrition Society*, **49**, 423–36.

Fuller, N. J., Jebb, S. A., Laskey, M. A., Coward, W. A. & Elia, M. (1992). Four-component model for the assessment of body composition in humans: comparison with alternative methods, and evaluation of density and hydration of fat-free mass. *Clinical Science*, **82**, 687–93.

Garrow, J. S. (1974). *Energy Balance and Obesity in Man*. Amsterdam: North-Holland Publishing Company.

Garrow, J. S., Stalley, S., Diethelm, R., Pittet, P., Hesp, R. & Halliday, D. (1979). A new method for measuring the body density of obese adults. *British Journal of Nutrition*, **42**, 173–83.

Girandola, R. N., Wiswell, R. A. & Romero, G. (1977). Body composition changes resulting from fluid ingestion and dehydration. *Research Quarterly in Exercise and Sports*, **48**, 299–303.

Gleichauf, C. N. & Roe, D. A. (1989). The menstrual cycle's effect on the reliability of bioimpedance measurements for assessing body composition. *American Journal of Clinical Nutrition*, **50**, 903–7.

Going, S. B., Massett, M. P., Hall, M. C., Bare, L. A., Root, P. A., Williams, D. P. & Lohman, T. G. (1993). Detection of small changes in body composition by dual-energy X-ray absorptiometry. *American Journal of Clinical Nutrition*, **57**, 845–50.

Grande, F. (1961). Nutrition and energy balance in body composition studies. In *Techniques for Measuring Body Composition*, ed. J. Borzek and A. Henschel, pp. 168–87. Washington, DC: National Academy of Sciences – National Research Council.

Graves, J. E., Pollock, M. L., Colvin, A. B., Van Loan, M. & Lohman, T. G. (1989). Comparison of different bioelectrical impedance analysers in the prediction of body composition. *American Journal of Human Biology*, **1**, 603–11.

Heitmann, B. L. (1990). Evaluation of body fat estimated from body mass index, skinfolds and impedance, a comparative study. *European Journal of Clinical Nutrition*, **44**, 831–7.

Heitmann, B. L. (1994). Impedance: a valid method in assessment of body composition. *European Journal of Clinical Nutrition*, **48**, 228–40.

Hortobagyi, T., Israel, R. G., Houmard, J. A., McCammon, M. R. & O'Brien, K. F. (1991). Comparison of body composition assessment by hydrodensitometry, skinfolds, and multiple site near-infrared spectrophotometry. *European Journal of Clinical Nutrition*, **46**, 205–11.

Israel, R. G., Houmard, J. A., O'Brien, K. F., McCammon, M. R., Zamora, B. S. & Eaton, A. W. (1989). Validity of a near-infrared spectrophotometry device for estimating human body composition. *Research Quarterly in Exercise and Sports*, **60**, 379–83.

Jackson, A. S. & Pollock, M. L. (1978). Generalised equations for predicting body density of men. *British Journal of Nutrition*, **40**, 497–504.

Jackson, A. S., Pollock, M. L. & Ward, A. (1980). Generalised equations for predicting body density of women. *Medicine and Science in Sports and Exercise*, **12**, 175–82.

James, W. P. T., Mascie-Taylor, C. G. N., Norgan, N. G., Bistrian, B. R., Shetty, P. S. & Ferro-Luzzi, A. (1995). The value of arm circumference measurements in assessing chronic energy deficiency in third world adults. *European Journal of Clinical Nutrition*, **48**, 883–94.

Jebb, S. A. & Elia, M. (1991). Assessment of changes in total body water in patients undergoing renal dialysis using bioelectrical impedance analysis. *Clinical Science*, **10**, 81–4.

Jebb, S. A. & Elia, M. (1993). Techniques for the measurement of body composition: a practical guide. *International Journal of Obesity*, **17**, 611–21.

Jebb, S. A., Murgatroyd, P. R., Goldberg, G. R., Prentice, A. M. & Coward, W. A.

(1993). *In vivo* measurement of changes in body composition: description of methods and their validation against 12-d continuous whole-body calorimetry. *American Journal of Clinical Nutrition*, **58**, 455–62.

Johansson, A. G., Forslund, A., Sjodin, A., Mallmin, H., Hambraeus, L. & Ljunghall, S. (1993). Determination of body composition – a comparison of dual energy X-ray absorptiometry and hydrodensitometry. *American Journal of Clinical Nutrition*, **57**, 323–6.

Katch, F. I. & Spiak, D. L. (1984). Validity of the Mellits and Cheek method for body-fat estimation in relation to menstrual cycle status in athletes and non-athletes below 22 per cent fat. *Annals of Human Biology*, **11**, 389–96.

Kushner, R. E., Kunigk, A., Alspaugh, M., Andronis, P. T., Leitch, C. A. & Schoeller, D. A. (1990). Validation of bioelectrical impedance analysis as a measurement of change in body composition in obesity. *American Journal of Clinical Nutrition*, **52**, 219–23.

Kushner, R. F., Schoeller, D. A., Fjeld, C. R. & Danford, L. (1992). Is the impedance index (ht^2/R) significant in predicting total body water. *American Journal of Clinical Nutrition*, **56**, 385–9.

Kvist, H., Chowdhury, B., Grangard, U., Tylen, U. & Sjostrom, L. (1988). Total and visceral adipose tissue volumes derived from measurements with computed tomography in adult men and women: predictive equations. *American Journal of Clinical Nutrition*, **48**, 1351–61.

Lohman, T. G. (1981). Skinfolds and body density and their relation to body fatness. *Human Biology*, **53**, 181–225.

Lohman, T. G. (1992). *Advances in Body Composition Assessment.* Champaign, Ill: Human Kinetics Publishers.

Lohman, T. G., Roche, A. F. & Martorell, R. (eds.) (1988). *Anthropometric Standardisation Reference Manual.* Champaign, Ill; Human Kinetics Books.

Lukaski, H. C. (1987). Methods for the assessment of human body composition: traditional and new. *American Journal of Clinical Nutrition*, **46**, 537–56.

McLean, K. P. & Skinner, J. S. (1992). Validity of Futrex-5000 for body composition determination. *Medicine and Science in Sports and Exercise*, **24**, 253–8.

McNeil, G., Fowler, P. A., Maughan, R. J., McGaw, B. A., Fuller, M. F., Gvozdanovic, D. & Gvozdanovic, S. (1991). Body fat in lean and overweight women estimated by six methods. *British Journal of Nutrition*, **65**, 95–103.

Malina, R. M. (1987). Bioelectric methods for estimating body composition; an overview and discussion. *Human Biology*, **59**, 329–35.

Martin, A. D. & Drinkwater, D. T. (1991). Variability in the measures of body fat. Assumptions or technique? *Sports Medicine*, **11**, 277–88.

Mazess, R. B. (1991). Do bioimpedance changes reflect weight, not composition? *American Journal of Clinical Nutrition*, **51**, 178.

Norgan, N. G. (1990). Body Mass Index and body energy stores in developing countries. *European Journal of Clinical Nutrition*, **44** (Suppl 1), 79–84.

Norgan, N. G. (1991a). Anthropometric assessment of body fat and fatness. In *Anthropometric Assessment of Nutritional Status*, ed. J. H. Himes, pp. 197–212. New York: A. R. Liss, Inc.

Norgan, N. G. (1991b). Densitometry. In *Nutritional Status Assessment. A Manual for Population Assessment*, ed. F. Fidanza, pp. 63–70. London: Chapman & Hall.

Norgan, N. G. (1992). Maternal body composition: methods for measuring short term changes. *Journal of Biosocial Science*, **24**, 367–77.

Norgan, N. G. (1994*a*). Population differences in body composition in relation to the Body Mass Index. *European Journal of Clinical Nutrition*, **48** (Suppl 3), S10–S27.

Norgan, N. G. (1994*b*). Relative sitting height and the interpretation of the body mass index. *Annals of Human Biology*, **21**, 79–82.

Norgan, N. G. (1995). The assessment of the body composition of populations. In *Body Composition Techniques and Assessment in Health and Disease*, ed. P. S. W. Davies and T. J. Cole, p. 195–221. Cambridge: Cambridge University Press.

Norgan, N. G. & Ferro-Luzzi, A. (1985). The estimation of body density in men: are general equations general? *Annals of Human Biology*, **12**, 1–15.

Oppliger, R. A., Looner, M. A. & Tipton, C. M. (1987). Reliability of hydrostatic weighing and skinfold measurements of body composition using a generalizability study. *Human Biology*, **59**, 77–96.

Oppliger, R. A. & Spray, J. A. (1987). Skinfold measurement variability in body density prediction. *Research Quarterly in Exercise and Sports*, **58**, 178–83.

Panter-Brick, C., Lotstein, D. S. & Ellison, P. T. (1993). Seasonality of reproductive function and weight loss in rural Nepali women. *Human Reproduction*, **8**, 684–90.

Patterson, P. (1992). The validity generalisation of skinfolds as measures of body density. *American Journal of Human Biology*, **4**, 115–24.

Pierson, R. N., Wang, J., Heymsfield, S. B., Russell-Aulet, M., Mazariegos, M., Tierney, M., Smith, R., Thornton, J. C., Kehayias, J., Weber, D. A. & Dilmanian, F. A. (1991). Measuring body fat: calibrating the rulers. Intermethod comparisons in 389 normal Caucasian subjects. *American Journal of Physiology*, **261** (Endocrinol. Metab.), E103–8.

Prentice, A. M., Goldberg, G. R., Jebb, S. A., Black, A. E., Murgatroyd, P. R. & Diaz, E. O. (1991). Physiological responses to slimming. *Proceedings of the Nutrition Society*, **50**, 441–58.

Pullicino, E., Coward, W. A., Stubbs, R. J. & Elia, M. (1990). Bedside and field methods for assessing body composition: comparison with the deuterium oxide dilution technique. *European Journal of Clinical Nutrition*, **44**, 753–62.

Reeves, J. (1979). Estimating fatness. *Science*, **204**, 881.

Rising, R., Swinburn, B., Larson, K. & Ravussin, E. (1991). Body composition in Pima Indians: validation of bioelectrical resistance. *American Journal of Clinical Nutrition*, **53**, 594–8.

Robinson, M. F. & Watson, P. E. (1965). Day-to-day variations in body weight of young women. *British Journal of Nutrition*, **19**, 225–35.

Ross, R., Shaw, K. D., Martel, Y., de Guise, J. & Avruch, L. (1993). Adipose tissue distribution measures by magnetic resonance imaging in obese women. *American Journal of Clinical Nutrition*, **57**, 470–5.

Roubenoff, R., Kehsyias, J. J., Dawson-Hughes, B. & Heymsfield, S. B. (1993). Use of dual-energy X-ray absorptiometry in body-composition studies: not yet a 'gold standard'. *American Journal of Clinical Nutrition*, **58**, 589–91.

Sanborn, C. F., Albrecht, B. H. & Wagner, W. W. (1987). Athletic amenorrhea:

lack of an association with body fat. *Medicine and Science in Sports and Exercise*, **19**, 207–12.

Segal, K. R., Van Loan, M., Fitzgerald, P. I., Hodgdon, J. A. & Van Itallie, T. B. (1988). Lean body mass estimation by bioelectrical impedance analysis: a four-site cross-validation study. *American Journal of Clinical Nutrition*, **47**, 7–14.

Shephard, R. J. (1991). *Body Composition in Biological Anthropology*. Cambridge: Cambridge University Press.

Shetty, P. S. & James, W. P. T. (1994). *Body Mass Index: A Measure of Chronic Energy Deficiency in Adults*. Rome: Food and Agricultural Organisation.

Sinning, W. E. & Wilson, J. R. (1984). Validity of 'generalised' equations for body composition analysis in women athletes. *Research Quarterly in Exercise and Sports*, **55**, 153–60.

Siri, W. S. (1961). Body composition from fluid spaces and density: analysis of methods. In *Techniques for Measuring Body Composition*, ed. J. Brozek and A. Henschel, pp. 223–44. Washington DC: National Academy of Sciences – National Research Council.

Sohlstrom, A., Wahlund, L.-O. & Forsum, E. (1993). Adipose tissue distribution as assessed by magnetic resonance imaging and total body fat by magnetic resonance imaging, underwater weighing, and body water dilution in healthy women. *American Journal of Clinical Nutrition*, **58**, 830–8.

Svodoba, M. D. & Query, L. M. (1984). Hydrostatic weighing throughout the menstrual cycle. In *Perspectives in Kinanthropometry*, ed. Olympic Scientific Congress, pp. 245–50. Champaign, Ill: Human Kinetics Publishers.

Thorland, W. G., Johnson, G. O., Tharp, G. D., Fagot, T. G. & Hammer, R. W. (1984). Validity of anthropometric equations for the estimation of body density in adolescent athletes. *Medicine and Science in Sports and Exercise*, **16**, 77–81.

Ulijaszek, S. J., Lourie, J. A., Taufa, T. & Pumuye, A. (1989). The Ok Tedi health and Nutrition Project, Papua New Guinea: adult physique of three populations in the North Fly region. *Annals of Human Biology*, **16**, 61–74.

Valdez, R., Seidell, J. C., Ahn, Y. I. & Weiss, K. M. (1993). A new index of abdominal obesity as an indicator of risk for cardio-vascular disease. A cross population study. *International Journal of Obesity*, **17**, 77–82.

van der Kooy, K. & Seidell, J. C. (1993). Techniques for the measurement of visceral fat: a practical guide. *International Journal of Obesity*, **17**, 187–96.

Van Loan, M. D. & Mayclin, P. L. (1992). Body composition assessment: dual-energy X-ray absorptiometry (DEXA) compared to reference methods. *European Journal of Clinical Nutrition*, **46**, 125–30.

van Raaij, J. M. A., Peek, M. E. M., Vermaat-Miedema, S. H., Schonk, C. M. & Hautvast, J. G. A. J. (1988). New equations for estimating body fat mass in pregnancy from body density or total body water. *American Journal of Clinical Nutrition*, **48**, 24–9.

Watson, P. E. & Robinson, M. F. (1965). Variations in body-weight of young women during the menstrual cycle. *British Journal of Nutrition*, **19**, 237–48.

Webber, J. & Macdonald, I. A. (1992). The changes in body weight and in body composition, estimated using bio-electrical impedance analysis over 72 h of fasting; comparisons with estimates based on energy expenditure measured by indirect calorimetry. *Proceedings of the Nutrition Society*, **52**, 251A.

Weiner, J. S. & Lourie, J. A. (eds.). (1981). *Practical Human Biology*. London: Academic Press.

Wong, W. W., Butte, N. F., O'Brian Smith, E., Garza, C. & Klein, P. D. (1989). Body composition of lactating women by anthropometry and deuterium dilution. *British Journal of Nutrition*, **61**, 25–33.

11 Breast-feeding practices and other metabolic loads affecting human reproduction

P. G. LUNN

Introduction

Few would dispute that, throughout the animal kingdom, successful species need to be able to increase their numbers when environmental conditions are favourable. The converse of this, that reproduction should be restricted in situations when a successful outcome is unlikely is also a clear evolutionary advantage, particularly for altricial species such as man with his long reproductive cycle and low rate of reproduction. Food supply is arguably the most potent evolutionary pressure, so some relationship between nutrition and reproduction must surely be expected. However, just how closely these factors are integrated, the mechanisms of the association and the overall impact of the relationship on lifetime fertility remain subjects of considerable controversy.

It is now several years since the concept of the baby in the driving seat was proposed in an attempt to explain some of the substantial variability observed in the duration of or lactational amenorrhoea (Lunn, 1985). In particular, the theory was used to explain the data obtained from a series of longitudinal studies into lactation–nutrition–fecundity interactions in a rural Gambian village where the length of post-partum infertility appeared to be sensitive to maternal nutritional adequacy (Lunn et al., 1980, 1984). It was suggested that the link between maternal nutritional status and fecundity was indirect and depended heavily on breast-feeding behaviour of individual mother–child dyads. The belief that by varying his suckling behaviour, the breast-feeding infant could drive maternal metabolism in a way most beneficial to himself, i.e. to increase milk production and to inhibit ovarian activity (so preventing the nutritional stress of a further pregnancy), appeared to explain most of the observations reported at the time. The alternative hypothesis (Frisch, 1984, 1985), that the duration of post-partum infecundity was a consequence of body weight and/or of fat

195

loss resulting from the energy cost of lactation could not be confirmed. Although small seasonal variation in body weight and composition were observed, they were far less than might have been expected on theoretical grounds, and certainly smaller than demanded by the Frisch theory (Frisch, 1985; Lunn, 1988).

Since that time, a wealth of new information has been generated and this review will examine the present day viability of these earlier concepts in the light of this more recent knowledge.

Energy requirements and metabolism during lactation

The energy cost of human lactation is generally considered to represent a substantial drain on maternal nutritional metabolism, but when compared with the situation facing other animals, these costs appear relatively small. A rat with 6–8 pups for example, needs to increase food consumption some 3–4 times and the gastrointestinal tract undergoes major modifications to deal with the extra load. Similarly, a sheep with two lambs needs a threefold rise in available energy (Prentice & Whitehead, 1987). In contrast, the most recent WHO/FAO estimates (WHO, 1985) indicate that an extra 2100 kJ/d is needed for a successful lactation and this represents only a 20–25% increase in maternal energy. Nevertheless, in developing countries with limited food availability, finding an extra 2100 kJ/d can be major problem. There is no doubt that the majority of women in developing countries somehow do manage to produce substantial amounts of milk (Prentice *et al.*, 1986), frequently for up to two years or more despite estimates of food consumption being well below expected levels. Moreover, most are able to do this without a significant reduction in their body weight. This raises the question of whether energy saving adaptations exist which help the maternal metabolism to accommodate the cost of lactation (Prentice & Prentice, 1990; Coward, Goldberg & Prentice, 1992).

In theory, the possible sources of energy for lactation are:

- Increased food consumption
- Decreased physical activity
- Reduced basal metabolic rate (BMR) and/or diet induced thermogenesis (DIT)
- Utilization of body fat stores.

and for balance to be maintained:

$$\text{Intake} = \text{BMR} + \text{Activity energy} + \text{milk energy} \pm \text{stored energy}$$

Table 11.1. *Energy adaptation to lactation (UK women)*

	Pre-pregnancy	Lactation
Body weight	57.1 kg	58.8 kg
Milk output (796 g/d)		2228 kJ/d
Basal metabolic rate (BMR)	5858 kJ/d	5787 kJ/d
Total energy expenditure (TEE)	9775 kJ/d	8955 kJ/d
Activity energy (TEE–BMR)	3917 kJ/d	3168 kJ/d
Activity index (TEE/BMR)	1.69	1.55
Energy intake	9422 kJ/d	10 881 kJ/d

Data from Goldberg *et al.* (1991).

One of the most complete investigations into maternal energy metabolism during lactation has been published recently by Goldberg *et al.* (1991). With the aid of doubly labelled water, most of the components of this equation were measured in well nourished breast-feeding women to establish how they had accommodated the increased nutritional requirements during the first three months of lactation (Table 11.1). Measurements showed that maternal body weight was higher than pre-pregnancy values by 1.71 kg, and that the body contained a greater proportion of fat. There was however no net change in body weight or fat content during the three-month study. Overall milk output was on average 796 g/d and contained 2233 kJ/d. Estimates of BMR showed that compared to pre-pregnant values there was a small, though not significant reduction. This is a rather surprising finding as this component of the equation would have been expected to increase as it incorporates the energy cost of milk synthesis, generally estimated to be about 20% of milk energy (Prentice & Prentice, 1990). Clearly, there must have been some metabolic adjustment to accommodate this cost but it may be that the explanation is simply that the energy released during milk synthesis was used to help maintain body temperature so that less thermogenic metabolism was required (Roberts & Coward, 1985). Apart from this small saving however, there was no evidence of a metabolic response which would markedly offset the increased nutritional stress. Total energy expenditure (TEE) of the women was nevertheless lower (8955 kJ/d) than pre-pregnancy values (9775 kJ/d), and subtracting the energy cost of BMR indicated that the energy had been conserved by a reduction of about 20% in physical activity. This was confirmed by the TEE:BMR ratio, often used as an index of activity, which had fallen from 1.69 to 1.55. The most substantial change recorded however, was in energy intake which had increased by about 16% during lactation from 9422 to 10 881 kJ/d.

Table 11.2. *Source of energy for milk production (UK women)*

Increase in energy intake	1459 kJ/d	(349 kCal/d)	62%
Decreased activity	820 kJ/d	(196 kCal/d)	35%
Decrease in BMR	71 kJ/d	(17 kCal/d)	3%
Total	2350 kJ/d	(562 kCal/d)	
Daily milk production was 796 g containing 2228 kJ (533 kCal)			

Data from Goldberg *et al.* (1991).

The contributions of these changes to the energy cost of milk production are summarised in Table 11.2. It shows that almost two-thirds of the cost of lactation had been met by increased food consumption and about one-third by a reduction in physical activity. Similar results have been reported from elsewhere (Illingworth *et al.*, 1986; Van Raaij *et al.*, 1991). To these values might be added a saving of about 200 kJ/d due to a reduction in diet-induced thermogenesis (Illingworth *et al.*, 1986), a component not measured in the Goldberg *et al.* (1991) study. It is important, however, to emphasize that these data are from well-nourished British women who could easily increase their food intake, but whose capacity for activity reduction was far more limited than, for example, women in the developing world who invariably have a far heavier workload. Unfortunately, no such complete investigation has yet been performed in a developing country although there are several reports in which some components of the system have been assessed.

Reported energy intakes from different third world countries have frequently been well below WHO/FAO recommendations and appear to be incompatible with a successful lactation unless major energy saving adaptations have developed in the mothers (Prentice, 1980). Recent studies in the Gambia, however, suggest that such metabolic alterations do not occur. Basal metabolic rates in lactating Gambian mothers have been found to be very similar to those of well-nourished British counterparts (Coward *et al.*, 1992) and exhibited only a small reduction, of about 5%, when compared with non-pregnant non-lactating Gambians (Lawrence & Whitehead, 1988). This is far less than the saving which would be necessary to achieve balance on the basis of their measured energy intakes. Moreover, their mean total energy expenditure was found to be 10 495 kJ/d which is similar to and perhaps even higher than values obtained in western mothers (Table 11.3). Subtraction of BMR from total expenditure showed the Gambians to have an extremely high energy expenditure (5103 kJ/d) on physical activity, presumably as a result of their heavy workload (Singh *et al.*, 1989).

Table 11.3. *Energy adaptation to lactation in Cambridge and Gambian women*

	Cambridge (UK)	The Gambia
Body weight	58.8 kg	49.5 kg
Milk output	2228 kJ/d	2100 kJ/d
Basal metabolic rate (BMR)	5787 kJ/d	5392 kJ/d
Total energy expenditure (TEE)	8955 kJ/d	10 495 kJ/d
Activity energy (TEE–BMR)	3168 kJ/d	5103 kJ/d
Activity index (TEE/BMR)	1.55	1.97
Energy intake	10 881 kJ/d	(?)

Data from Goldberg *et al.* (1991) and Singh *et al.* (1989).

In theory, it is possible that a substantial mobilization of body fat might have occurred to offset some of this high expenditure and indeed in this particular study which was carried out in the hard-working rainy season, some fat loss was recorded. In The Gambia, however, the role of maternal body fat as an energy source for lactation does not hold. Fat stores in the population in general vary with season, and lactating women are not exempt from this trend. During the dry season, when work loads are lower they gain fat whatever their stage of lactation, whereas in the wet, hard working season they lose it. Whether fat is gained or lost thus reflects the season and not lactation (Prentice *et al.*, 1981). The whole concept of fat stores being used to supplement lactation may not be valid in developing countries (Prentice & Prentice, 1990).

Thus the overall maximum energy savings available to third world women are probably as shown in Table 11.4. From this data it does appear feasible that the total cost of lactation in such women could be accommodated without increased food consumption. However, such a possibility would depend almost entirely on a reduction in physical activity, which in the third world means a reduction in workload. In the Gambia, maternal workload was only markedly diminished during the first few weeks of lactation and after three months no decrease could be detected (Lawrence & Whitehead, 1988). By this time the women were back to working their usual long hours in the fields, an observation confirmed by a high TEE:BMR activity ratio close to 2.0 (Singh *et al.*, 1989). This early return to heavy work seems to be the norm in most developing countries, yet it is here where substantial savings in energy could be made (Guillermo-Tuazon *et al.*, 1992). Such considerations, however, must be balanced against the fact that, in many cases, the work needs to be done to ensure a food supply and thus survival of the family.

Table 11.4. *Probable maximum energy savings in third-world women*

Reduction in BMR	saves 500 kJ/d
Reduction in diet-induced thermogenesis	saves 200 kJ/d
Reduced activity (workload)	could save 500–1500 kJ/d
	Total 1200–2220 kJ/d
	= 287–526 kCal/d
Energy content of 750 ml of milk	2100 kJ (502 kCal)

The doubly labelled water technique has shown total energy expenditure of lactating Gambian women to be far higher than would be possible from the estimates of dietary intake (Coward *et al.*, 1992). In addition, the BMR and fat mobilisation measurements have demonstrated that the potential for metabolic adaptation in lactating women is extremely limited. It therefore seems certain that estimates of maternal food consumption in The Gambia and probably those from other countries must be highly inaccurate as they are simply incompatible with energy utilization. Such serious doubts about the validity of energy intake measurements make any inferences about the increment in energy intake accompanying lactation virtually impossible (Prentice & Prentice, 1990).

These studies suggest that food consumption of third world lactating mothers is almost certainly significantly greater than was previously believed. The nutritional stress which these women suffer appears to occur primarily as a result of their excessive workload whilst lactating rather than to an inadequate supply of food. Consequently, studies of athletes in western communities probably provide a more valuable insight into questions concerning fertility in developing countries than has been previously thought.

Nutrition and fertility

The nutritional requirements of the mother–child dyad are clearly a combination of the needs of the mother and her infant. At any time, however, their levels of hunger or satiety may vary independently, so it is perhaps not surprising that each should have their own individual mechanism for relating their personal nutritional status to reproductive activity. Thus the two theories which been proposed in an attempt to explain the links between nutrition and reproduction should be viewed as being complimentary rather than contradictory. Whilst maternally generated mechanisms are aimed at protecting the mother's nutritional status, in the suckling reflex mechanism, the child is protecting his own present and future nutrient supply.

Maternal mechanisms, nutritional status

Some 20 years have passed since Frisch and McArthur (1974) introduced their hypothesis relating to the onset of menstrual cycling to nutritional status. The exact form of the hypothesis has been modified and extended since that time and now states that a minimum body weight and/or fat content is necessary for women to achieve and to maintain menstrual cycling (Frisch, 1987). Although there has never been any doubt that reproductive function was sensitive to severe nutritional stress, e.g. in famine (Stein & Susser, 1978) and in anorexia nervosa (Nillius, 1983), attempts to apply the Frisch hypothesis to women with mild or moderate malnutrition or to lactating mothers has failed to give convincing results. More support for the theory, however, has come from investigations of amenorrhoea in female athletes.

In 1978, Feicht *et al.* reported a high prevalence of amenorrhoea (defined as three periods or less per year) in female track and field athletes. They also found that the frequency of menstrual dysfunction varied from 6 to 43% depending on their weekly training mileage. Subsequently, these findings have been confirmed in many, though by no means all, studies of women over a wide range of different sporting activities (see Cumming, Wheeler & Harber, 1994). Several reports have concluded that amenorrhoeic women were lighter, leaner and had lost more body weight than their normally menstruating counterparts (Speroff & Redwine, 1979; Dale, Getlach & Wilhite, 1979; Schwartz *et al.*, 1981; Carlberg *et al.*, 1983). Others, however, have failed to find such relationships (Wakat, Sweeney & Rogol, 1978; Shangold & Levine, 1982; Sanborn, Albrecht & Wagner, 1987). Although it must be accepted that the various techniques used to assess body fat are all indirect and do not always agree in absolute terms (Cumming & Rebar, 1984), it does seem certain that low body fat is not invariably associated with menstrual irregularities. Menstrual dysfunction appears to be more common in sports activities in which a low body fat is beneficial for success (Sanborn, Martin & Wagner, 1982; Brooks-Gunn, Burrow & Warren, 1988) but even here, irregular menses are not invariably related to low fat content (Cumming *et al.*, 1994). Nevertheless, there is equally no doubt that low body weight, dieting, exercise, loss of body weight and fat are associated with amenorrhoea and oligomenorrhoea and such relationships clearly necessitate further explanation.

In 1987, Frisch answered one of the more persistent criticisms of her hypothesis, i.e. that no physiological mechanism linking body fat with ovarian activity was known. She identified four separate mechanisms by which metabolic activity within adipose tissue might modify the hypothalamic–pituitary–ovarian axis (Frisch, 1987). Firstly, fat tissue may enhance the conversion of androgens to oestrogen (Siiteri & MacDonald, 1973). Certainly this tissue is one of the most important extragonadal sources of oestrogens in pre-menopausal women and the major source post-menopause (Siiteri, 1981). Secondly, it appears that the degree of fatness influences the direction of oestrogen metabolism to either more or less active forms (Fishman, Boyar & Hellman, 1975). Lean women athletes have elevated blood concentrations of the relatively inactive 2-hydroxylated oestrogens (Snow, Barberi & Frisch, 1989; Frisch et al., 1993) whereas obese women have more of the highly active 16-hydroxylated form (Schneider et al., 1983). Reductions in blood levels of sex-hormone binding globulin have also been observed in fat women and girls (Siiteri, 1981; Apter et al., 1984). This would lead to an increase in the concentration of free hormone which is generally believed to be the active form. Finally, adipose tissue is known to store oestrogens, but how this action affects circulating levels is not clear (Kaku, 1969). Although there is no doubt that these differences in metabolism between fat and thin women exist, there is no evidence available to show whether these proposed mechanisms do indeed play a role in the control of ovarian activity.

There is, however, an alternative explanation for the many of the observations reported on the relationship between body-weight or fatness and menstrual dysfunction. The advent of simple techniques for the measurement of the gonadotrophic hormones luteinising hormone (LH) and follicle stimulating hormone (FSH), followed by assays for oestrogens and progestogens in saliva and urine has led to the realization that a sensitive mechanism linking maternal nutritional status to ovarian activity does exist. Initially such hormone measurements were used to examine the cause of oligomenorrhoea and amenorrhoea in female athletes (Dale et al., 1979; Schwartz et al., 1981) and ballet dancers (Frisch, Wyshak & Vincent, 1980; Warren, 1980). Reductions in the levels of oestrogens and progestogens were found and it is now generally accepted that these low values reflect diminished LH and FSH production by the pituitary (Cumming et al., 1985a,b). This, in turn, appears to result from a down regulation of the hypothalamic gonadotrophin pulse generator (Veldhuis et al., 1985; Loucks et al., 1989). The severity of the effects seems to be related to the duration and intensity of training, particularly if it is accompanied by a self-imposed dietary restraint. Such women are frequently underweight and have a low body fat content (Tremblay *et al.*, 1990), thus these more severe cases provide an endocrinological explanation of the Frisch hypothesis.

However, steroid hormone measurements provide a far more sensitive index of fertility than the presence or absence of menstruation and there is now considerable evidence that hormone values are also reduced in response to levels of diet and/or exercise which result in minimal or zero weight loss (Ellison & Lager, 1986). The study of Bullen *et al.* (1985) was particularly important as the subjects were previously sedentary women and not athletes who for reasons of physique could be biased towards lower ovarian activity (Malina, 1983). The participants underwent a two month period of strenuous exercise, running up to 10 miles per day, with a dietary regime which allowed one group to maintain body weight whilst the other lost weight at an average 0.45 kg per week. Fifteen out of sixteen in the weight loss group showed some disruption of normal ovarian function, but so did 75% of the women who did not lose weight. Abnormalities included delayed menses, shortened or deficient luteal phases and failure of the LH surge. All the women had re-established normal cycles 12 weeks after finishing the investigation highlighting the reversible nature of the changes. In another study with less severe exercise, recreational runners averaging only 12.5 miles per week were found to have lower and shorter profiles of salivary progesterone than sedentary controls despite having menstrual cycles of similar length and a slightly heavier body weight (Ellison & Lager, 1986; Lager & Ellison, 1987). Although the endocrine changes were less severe than in the previous study, they nevertheless indicated an increased incidence of ovulatory failure and/or inadequate luteal cycles even in women whose menstrual cycling was normal.

Of particular importance is the fact that these effects of exercise can be ameliorated by raised dietary intake (Bullen *et al.*, 1985). Cameron *et al.* (1990) have reported that amenorrhoea induced by chronic exercise in monkeys could be reversed by increased food intake. Other work with primates has shown that LH pulsatility responded to dietary changes within 24 hours long before there was any substantial weight loss (Cameron & Nosbisch, 1991). It has not been confirmed, however, whether these acute changes can be extrapolated to the long term situation observed in continually training athletes and clearly more research is required in this area. Nevertheless, it is undoubtedly true that diet and exercise are on opposite sides of the nutritional stress equation.

Ellison, Peacock and Lager (1989) have obtained an entirely analogous set of results from their studies of Lese women in the Ituri Forest region of Zaire. These women differ from their western counterparts in that their high energy expenditure on activity and their poor diet are matters of survival and not choice. Salivary progesterone measurements in women aged between 23 and 50 showed the percentage of ovulatory cycles to be only about 60%, compared to almost 100% in well-nourished American women (Ellison, 1991). In addition, over a 4-month period of food scarcity the percentage of ovulatory cycles decreased from 63% to 30% although women lost less than 1 kg of body weight. Similar data has been obtained from studies of the Tamang of Nepal and rural agricultural workers in Poland. In the Tamang, seasonal changes in workload were reflected in salivary progesterone concentrations with particularly low levels occurring in those who lost weight during the period of heavy labour (Panter-Brick, Lotstein & Ellison, 1993). The endocrine changes were mirrored by a decrease in conception at this time. In Poland, progesterone values were lower in the women with heavier workloads even though no differences could be detected in body weight or in menstrual cycling (Jasienska & Ellison, 1993).

Nutritional status and nutritional stress

The common denominator in all the reports discussed above and many others is that reproductive abnormalities are more likely to occur in situations of nutritional stress. This can be brought about either by increased exercise or decreased food intake and a combination of both leads to more severe effects, but how is nutritional stress related to nutritional status, and how can it be assessed?

Apart from the use of special tests to detect specific nutrient deficiencies, nutritional status has invariably been assessed by anthropometric measurement and/or indirect estimates of body-fat content. Yet it is clear from the studies described above that moderate levels of exercise can quickly lead to a down regulation of reproductive activity long before anthropometric changes can be detected. In other, longer term studies, reductions in reproductive hormone concentration occurred even though anthropometric losses were either zero or far smaller than those demanded by the Frisch hypothesis (Frisch, 1987). Nevertheless, when body weight was compromised, the reduction in hormone concentrations became more marked and there was an increased likelihood of reduced fecundity and even amenorrhoea (Loucks et al., 1989; Ellison, 1991). These results can be taken to imply that although a relationship between nutrition and fecundity exists, anthropometric measurements such as bodyweight, BMI or estimates of body fat are simply too insensitive to assess the true nutritional stress of an individual, and are particularly inadequate when the subject is close to balance.

Although the concept of 'nutritional stress' is easily understood, an exact definition of the term is far more difficult. Clearly, it is related to nutritional balance but probably not in a straightforward way. Moreover, it should be appreciated that although this discussion considers only energy status, similar regard should be given to each of the other essential macro- and micro-nutrients. When the balance between energy intake and expenditure is negative and tissues are being mobilized to make up the shortfall, the subject is clearly nutritionally stressed and reductions in reproductive indices ensue relative to the severity of the imbalance. If this level of stress is prolonged, body-weight and fat loss will eventually become measurable. However marked differences in hormone levels are also apparent between sedentary and exercising women even when both are in energy balance. In this circumstance ovarian activity appears to decline with increasing physical energy expenditure. Thus any definition of nutritional stress must incorporate the issue of high energy turnover or flux as well as energy balance. Moreover, if it is accepted that increased energy turnover can reduce fecundity in the absence of weight loss, the increased turnover and nutritional stress imposed by lactation and the contribution of this mechanism to lactational amenorrhoea also must be considered.

How a situation of maternal nutritional stress might be relayed to the hypothalamus and thus affect ovarian activity remains to be elucidated (See Thalabard, this volume). The insensitivity of anthropometric assessment of nutritional stress was again demonstrated in the Gambian studies when the addition of some 3000 kJ/d to the diet of lactating mothers had minimal effects on body weight or composition (Prentice *et al.*, 1983), yet large changes in fecundity were recorded (Lunn *et al.*, 1984). We can only assume that, apart from relieving the nutritional stress imposed by lactation, any extra energy was utilized in increased activity. This and other data suggest that some more dynamic index, related to energy turnover or balance or even the mechanisms responsible for partitioning nutrients within the body may be the controlling factor. The small reduction in BMR which has been reported in both amenorrhoeic athletes (Myerson *et al.*, 1991) and lactating mothers (Goldberg *et al.*, 1991) may be a reflection of this process and, perhaps, could itself elicit the effect. Thus the athletes drive for muscle activity (Fig. 11.1a) may put a stress on the body's internal energy balance which takes precedence over maintaining reproductive functions. Similarly, one could argue that the infant stimulated drive to milk production could take precedence over ovarian function (Fig. 11.1b) especially in the third world context where there is probably also a high

a) **Western athletes**

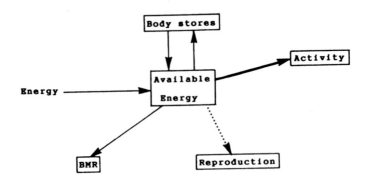

b) **Lactating Gambians**

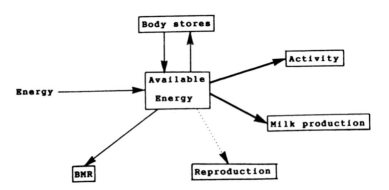

Fig. 11.1. Nutrient partitioning in a), training athletes and b) lactating third world mothers.

activity expenditure. Taken further still, this theory would predict that the nutritional stress of milk production itself could lead to lactational amenorrhoea irrespective of the suckling reflex arc. When the nutritional drive to milk production ceased, e.g. at weaning or infant death, nutritional stress would be eased and reproductive function could quickly return. Increasing food consumption in exercise schedules moves the hormone levels back towards normal, thus improving food supply to malnourished lactating women might be expected to result in a shorter period of lactational amenorrhoea. Could suckling intensity and plasma prolactin level simply be a proxies for the nutritional stress of milk production rather than determinants of ovarian inhibition? This is probably taking the argument too far, however, it would be extremely surprising if this mechanism did not interact in some way with suckling-induced subfecundity especially as both appear to act via the hypothalamic gonadotrophin pulse generator.

Infant mechanisms, the suckling reflex arc

The role of lactation and the importance of suckling in maintaining post-partum infecundity is well known and the endocrinological mechanisms involved have now been well characterized (McNeilly, 1993). There is little doubt that in non-contracepting populations this mechanism provides the greatest single contribution to birth spacing (Howie, 1993). The suckling reflex arc is initiated by sensors in the areolar region of the breast which have direct neural connections to the hypothalamus. The central feature of the mechanism is that while the infant is taking milk from his mother, the hypothalamus is stimulated and the ensuing release of prolactin from the pituitary promotes the synthesis of milk ready for the next feed. Until recently, the same increased prolactin concentrations brought about by suckling were also believed to directly inhibit ovarian activity and thus constitute the mechanism of lactational amenorrhoea (Delvoye *et al.*, 1978; Lunn, 1985). Now it seems that any direct effect of prolactin in suppressing ovarian activity is unlikely and that the impact of suckling stimulus on the ovaries is mediated via the hypothalamic gonadotrophin pulse generator (McNeilly *et al.*, 1994). Investigations into the nature of the ovarian inhibition and endocrine measurements have allowed better estimates of the true duration of lactational infecundity, and it is now clear that, because of anovular cycles or inadequate corpora luteal function, lactational subfecundity persists long after menstrual cycling has recommenced (Walker, Walker & Riad-Fahmy, 1984; Ellison, 1991). It seems likely that this extra period of subfecundity explains at least part of the so called 'waiting time' between resumption of menstruation and next conception (Delvoye, Delogne-Desnoeck & Robyn, 1980; Howie & McNeilly, 1982). Whether or not these mechanism are sensitive to the nutritional status of the mother and/or her child, however, remains a subject of controversy.

The 'baby in the driving seat' concept was introduced to explain the relationship observed between maternal nutrition and lactational infecundity in a series of longitudinal study conducted in a rural area of the Gambia (Lunn *et al.*, 1984; Lunn, 1985). In brief, these studies showed that when marginally nourished women were given a large dietary supplement, either during lactation alone or during both pregnancy and lactation, there was a substantial shortening of the duration of post-partum infecundity of up to 8 months and moreover, a reduced time to conception. These changes were associated with reductions in plasma prolactin concentration, and it was argued that the extra food had improved milk availability resulting in less vigorous suckling by the baby. The less intense suckling, had in turn, decreased ovarian suppression. It was suggested that in developing countries where maternal nutritional status was poor, a hungry baby, by suckling more frequently or for longer, could stimulate a greater release of prolactin which not only drove the maternal metabolism towards milk production but also strengthened the inhibitory effect on ovarian activity.

The more recent knowledge which disassociates prolactin from ovarian activity, however, makes this interpretation of the Gambian data far less secure. Whilst prolactin was believed to have the dual role of promoting milk synthesis and inhibiting reproductive activity, a close link between these two processes and maternal nutrition was assured. The current interpretation is that the degree of elevation of plasma prolactin is an indicator of suckling strength and thus the strength of the neuro-hypothalamic signal (McNeilly, 1993). Although this may be the case, it is quite possible that the response of the gonadotrophin pulse generator to the suckling stimulus may be greatly influenced by other inputs, for example the nutritional stress response (Fig. 11.2). Is there any evidence in published data that suggests such an interaction might occur? The question can be rephrased to ask if all the variation observed in the duration of lactational infecundity can be explained in terms of the suckling reflex arc. This is a far from easy question to answer because it is generally believed that the sensitivity of the suckling reflex mechanism allows for a near infinite variation in final neural signal to the hypothalamus (Lunn, 1992). However, there is no doubt that the efficacy of nursing in suppressing ovarian activity varies widely among individuals and populations and it has proven extremely difficult to designate a level of suckling activity which will ensure infecundity. Even though details of minimum suckling frequency and duration to maintain amenorrhoea had already been suggested (Howie & McNeilly, 1982; Anderson & Scholer, 1982; Stern et al., 1986) the Bellagio consensus recommendation (1988) was simply that full breast-feeding could be relied on to give a 98% protection against pregnancy, but only up to six months post-partum and only if the mother remained amenorrhoeic. Even during the first six months post-partum, these recommendations are less than 100% safe in well-fed westernized popula-tions but they are perhaps unnecessarily stringent for malnourished

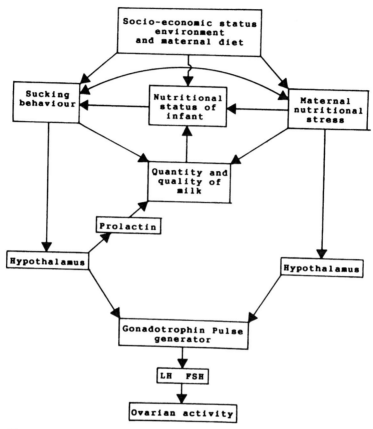

Fig. 11.2. Interrelationships between nutrition and fecundity in lactating women.

third-world women whose lactational amenorrhoea frequently lasts 18 months or longer despite the introduction of weaning foods at an early age. In contrast, Worthman *et al.* (1993) has described a surprisingly short period of lactational infecundity in Amele women of New Guinea and concluded that even when nursing is intensive and persistent, women of good nutritional status do not experience the same degree of ovarian suppression as do less well-nourished populations with similar nursing patterns. Small differences in breast-feeding patterns between the Tamang and Kami castes of Nepal seem unable to explain the higher plasma prolactin concentrations and much longer lactational amenorrhoea and birth intervals of the Tamang mothers (Panter-Brick, 1991; Stallings, Panter-Brick & Worthman, 1994). Diaz *et al.* (1991) have reported that women who experienced a short lactational amenorrhoea (less than 180 days) had lower plasma oestradiol and higher prolactin concentrations from one month post-partum than those who were amenorrheic for longer despite similar breast-feeding frequencies and duration. They explained their results by suggesting that the hypothalamic response to similar suckling stimuli varied between individuals. These observations clearly fall far short of proof but could suggest that the sensitivity of the hypothalamic gonadotrophin generator to suckling may be modulated by other mechanisms, one of which may be maternal nutritional stress.

To evaluate the relative importance of these two nutritional mechanisms in suppressing ovarian activity during lactation will require differential measurements of the response of the pulse generator to these and perhaps other, psychological? inputs. It is a clear challenge for the future.

Summary

Major advances in three areas have been described which have clarified some issues but added further complexities to others.

Measurements of energy expenditure of lactating mothers in the Gambia have radically changed our ideas of maternal energy balance in developing countries. The data has demonstrated that metabolic adaptations for conserving energy are minimal and that the level of energy expenditure was quite incompatible with estimates of energy intake. As the measured energy consumption of lactating Gambian mothers was similar to data from other developing countries, doubt must be expressed about the validity of dietary data elsewhere. It seems likely that in many populations, nutritional stress occurs primarily as a result of heavy work-loads rather than dietary inadequacy.

The ability to measure reproductive hormone levels in saliva and urine has led to a greater understanding of the relationship between maternal nutrition and fecundity. The mechanism is far more sensitive than suggested by older theories which related failure of menses to moderate or severe weight loss. Amenorrhoea can now be seen as the most severe stage of a graded continuum of ovarian suppression and thus fecundity, as nutritional stress increases. Likewise, the ability to detect these less severe stages of dysfunction has shown that reproductive activity is compromised at levels of nutritional stress well below that required for sustained weight loss. Our ability to quantify maternal nutritional stress and even an exact definition of the term remains problematical. Anthropometric measurements, even when collected longitudinally can give only a crude estimate of long term trends. Measurements of food intake must be viewed with suspicion and even then are probably of little use without concurrent estimations of energy expenditure. Questions of adaptation and training must be added to the equation as should the possibility of deficiencies of specific micronutrients. Overall, the evidence suggests that the reproductive system is sensitive to small changes in the difference between energy expenditure and intake, but how this information is transmitted to the pulse generator and how to measure the signal remain to be elucidated.

The significance of the suckling reflex arc in lactational infecundity has been reinforced by an increased understanding of the mechanisms involved. However, as both suckling and nutritional stress appear to operate through the hypothalamic gonadotrophin pulse generator, the way in which the two signals are integrated, particularly in terms of maternal and infant nutrition, represents a major challenge for future research.

Whatever such studies finally prove, from a nutrition–fecundity point of view, the overall effects of both ovarian inhibitory mechanisms are the same; nutritional stress in lactating mothers will lead to prolonged lactational amenorrhoea and subfecundity. Either a better diet or a reduction in maternal workload during lactation can be expected to shorten the duration of lactational infecundity. Whether the interaction is important in terms of lifetime fertility and population growth requires the use of more appropriate variables than have been investigated to date.

REFERENCES

Andersen, A. N. & Schloer, V. (1982). Influence of breast-feeding pattern on pituitary–ovarian axis of women in an industrialised community. *American Journal of Obstetrics and Gynaecology*, **143**, 673–7.
Apter, D., Bolton, N. J., Hammond, G. L. & Vihko, R. (1984). Serum sex hormone-binding globulin during puberty in girls and in different types of adolescent menstrual cycles. *Acta Endocrinologica*, **107**, 413–19.
Bellagio consensus statement (1988). Breastfeeding as a family planning method. *Lancet*, **ii**, 1204–5.
Brookes-Gunn, J., Burrow, D. & Warren, M. P. (1988). Attitudes towards eating and body weight in different groups of female athletes. *International Journal of Eating Disorders*, **7**, 749–57.
Bullen, B. A., Skrinar, G. S., Beitins, I. Z., von Mering, G., Turnbull, B. A. & McArthur, J. W. (1985). Induction of menstrual disorders by strenuous exercise in untrained women. *New England Journal of Medicine*, **312**, 1349–53.
Cameron, J. L. & Nosbisch, C. (1991). Suppression of pulsatile LH and testosterone secretion during short-term food restriction in adult male rhesus monkeys. (*Macaca mulatta*). *Endocrinology*, **128**, 1532–40.

212   P. G. Lunn

Cameron, J. L., Nosbisch, C., Helmreich, D. L. & Parfitt, D. B. (1990). Reversal of exercise-induced amenorrhoea in female cynomolgus monkeys. *Abstracts of the 72nd Annual Meeting of the Endocrine Society*, No. 1042.

Carlberg, K. A., Buckman, M. T., Peake, G. T. & Riedesel, M. L. (1983). Body composition of oligo/amenorrheic athletes. *Medicine and Science of Sports and Exercise*, 15, 215–17.

Coward, W. A., Goldberg, G. & Prentice, A. M. (1992). Energy balance in lactation. In *Mechanisms Regulating Lactation and Infant Nutrient Utilization*, ed. M. F. Picciano and B. Lönnerdal, pp. 65–76. New York: Wiley-Liss.

Cumming, C. D. & Rebar, R. W. (1984). Lack of consistency in the indirect methods of estimating body fat. *Fertility and Sterility*, 41, 739–42.

Cumming, C. D., Vickovic, M. M., Wall, S. R. & Fluker, M. R. (1985a). Defects in pulsatile LH release in normally menstruating runners. *Journal of Clinical Endocrinology and Metabolism*, 60, 810–12.

Cumming, C. D., Vickovic, M. M., Wall, S. R., Fluker, M. R. & Belcastro, A. N. (1985b). The effect of acute exercise on pulsatile release of luteinising hormone in women runners. *American Journal of Obstetrics and Gynecology*, 153, 482–5.

Cumming, C. D., Wheeler, G. D. & Harber, V. J. (1994). Physical activity, nutrition and reproduction. *Annals of the New York Academy of Sciences*, 709, 55–76.

Dale, E., Gerlach, D. H. & Wilhite, A. L. (1979). Menstrual dysfunction in distance runners. *Obstetrics and Gynecology*, 54, 47–53.

Delvoye, P., Demaegd, M., Uwayitu-Nyampeta & Robyn, C. (1978). Serum prolactin, gonadotropins and estradiol in menstruating and amenorrheic mothers during two years of lactation. *American Journal of Obstetrics and Gynecology*, 130, 635–9.

Delvoye, P., Delogne-Desnoeck, J. & Robyn, C. (1980). Long-lasting hyperprolactinaemia: evidence for anovulatory cycle and inadequate corpus luteum. *Clinical Endocrinology*, 13, 243–7.

Diaz, S., Cardenas, H., Brandeis, A., Miranda, P., Schiappa-casse, V., Salvatierra, A. M., Herreros, C., Seron-Ferre, M. & Croxatto, H. B. (1991). Early difference in the endocrine profile of long and short lactational amenorrhoea. *Journal of Clinical Endocrinology and Metabolism*, 72, 196–201.

Ellison, P. T. (1991). Reproductive ecology and human fertility. In *Applications of Biological Anthropology to Human Affairs*, ed. G. W. Lasker and C. G. N. Mascie-Taylor, pp. 14–54. Cambridge: Cambridge University Press.

Ellison, P. T. & Lager, C. (1986). Moderate recreational running is associated with lowered salivary progesterone profiles in women. *American Journal of Obstetrics and Gynecology*, 154, 1000–3.

Ellison, P. T., Peacock, N. A. & Lager, C. (1989). Ecology and ovarian function among Lese women of the Ituri Forest, Zaire. *American Journal of Physiological Anthropology*, 78, 519–26.

Feicht, C. B., Johnson, T. S., Martin, B. J., Sparkes, K. E. & Wagner, W. W. (1978). Secondary amenorrhoea in athletes. *Lancet*, ii, 1145.

Fishman, J., Boyar, R. M. & Hellman, L. (1975). Influence of body weight on estradiol metabolism in young women. *Journal of Clinical Endocrinology and Metabolism*, 41, 989–91.

Frisch, R. E. (1984). Body fat, puberty and fertility. *Biological Reviews*, 59, 161–88.

Frisch, R. E. (1985). Maternal nutrition and lactational amenorrhoea: perceiving the metabolic costs. In *Maternal Nutrition and Lactational Infertility*, ed. J. Dobbing, pp. 65–91. New York: Nestlé Nutrition, Vevey/Raven Press.

Frisch, R. E. (1987). Body fat, menarche, fitness and fertility. *Human Reproduction*, 2, 521–33.

Frisch, R. E. & McArthur, J. W. (1974). Menstrual cycles: fatness as a determinant of minimum weight for height necessary for their maintenance and onset. *Science*, 185, 949–51.

Frisch, R. E., Wyshak, G. & Vincent, L. (1980). Delayed menarche and amenorrhea of ballet dancers. *New England Journal of Medicine*, 303, 17–19.

Frisch, R. E., Snow, R. C., Johnson, L. A., Gerard, B., Barberi, R. & Rosen, B. (1993). Magnetic resonance imaging of overall and regional body fat, estrogen metabolism and ovulation of athletes compared to controls. *Journal of Clinical Endocrinology and Metabolism*, 77, 471–7.

Goldberg, G., Prentice, A. M., Coward, W. A., Davies, H. L., Murgatroyd, P. R., Sawyer, M. B., Ashford, J. & Black, A. E. (1991). Longitudinal assessment of the components of energy balance in well-nourished lactating women. *American Journal of Clinical Nutrition*, 54, 788–98.

Guillermo-Tuazon, M. A., Barba, C. V., Van Raaij, J. M. & Hautvast, J. G. (1992). Energy intake, energy expenditure, and body composition of poor rural Philippine women throughout the first 6 months of lactation. *American Journal of Clinical Nutrition*, 56, 874–80.

Howie, P. W. (1993). Natural regulation of fertility. *British Medical Bulletin*, 49, 182–99.

Howie, P. W. & McNeilly, A. S. (1982). Effect of breast feeding patterns on human birth intervals. *Journal of Reproduction and Fertility*, 65, 545–57.

Illingworth, P. J., Jung, R. T., Howie, P. W., Leslie, P. & Isles, T. E. (1986). Diminution in energy expenditure during lactation. *British Medical Journal*, 292, 437–41.

Jasienska, G. & Ellison, P. T. (1993). Heavy workload impairs ovarian function in Polish peasant women. *American Journal of Physiological Anthropology*, Supplement 16, 117–18.

Kaku, M. (1969). Disturbance of sexual function and adipose tissue of obese females. *Sanfujinka No Jissai*, (Tokyo), 18, 212–18.

Lager, C. & Ellison, P. T. (1987). Effects of moderate weight loss on ovulatory frequency and luteal function in adult women. *American Journal of Physiological Anthropology*, 72, 221–2.

Lawrence, M. & Whitehead, R. G. (1988). Physical activity and total energy expenditure of child-bearing Gambian village women. *European Journal of Clinical Nutrition*, 42, 145–60.

Loucks, A. B., Mortola, J. F., Girton, L. & Yen, S. S. C. (1989). Alterations in the hypothalamic–pituitary–ovarian and the hypothalamic–pituitary–adrenal axes in athletic women. *Journal of Clinical Endocrinology and Metabolism*, 68, 402–11.

Lunn, P. G. (1985). Maternal nutrition and lactational infertility: the baby in the driving seat. In *Maternal Nutrition and Lactational Infertility*, ed. J. Dobbing, pp. 41–64. New York: Nestlé Nutrition, Vevey/Raven Press.

Lunn, P. G. (1988). Malnutrition and fertility. In *Natural Human Fertility: Social and Biological Determinants*, ed. P. Diggory, M. Potts and S. Teper, pp. 135–52. Basingstoke: Eugenics Society, Macmillan Press.

Lunn, P. G. (1992). Breast-feeding patterns, maternal milk output and lactational infecundity. *Journal of Biosocial Science*, **24**, 317–24.

Lunn, P. G., Prentice, A. M., Austin, S. & Whitehead, R. G. (1980). Influence of maternal diet on plasma-prolactin levels during lactation. *Lancet*, **i**, 623–5.

Lunn, P. G., Austin, S., Prentice, A. M. & Whitehead, R. G. (1984). The effect of improved nutrition on plasma prolactin concentrations and postpartum infertility in lactating Gambian women. *American Journal of Clinical Nutrition*, **39**, 227–35.

Malina, R. M. (1983). Menarche in athletes: a synthesis and hypothesis. *Annals of Human Biology*, **10**, 1–24.

McNeilly, A. S. (1993). Lactational amenorrhoea. *Endocrinology and Metabolism Clinics of North America*, **22**, 59–73.

McNeilly, A. S., Tay, C. C. K. & Glasier, A. (1994). Physiological mechanisms underlying lactational amenorrhea. *Annals of the New York Academy of Sciences*, **709**, 145–55.

Myerson, M., Gutin, B., Warren, M. P., May, M. T., Contento, I., Lee, M., Pi-Sunyer, F. X., Pierson, R. N. & Brookes-Gunn, J. (1991). Resting metabolic weight and energy balance in amenorrheic and eumenorrheic runners. *Medicine and Science of Sports and Exercise*, **23**, 15–22.

Nillius, S. J. (1983). Weight and the menstrual cycle. In *Understanding Anorexia Nervosa and Bulimia*. Report of the Fourth Ross Conference on Medical Research, pp. 77–81. Columbus, Ohio: Ross Laboratories.

Panter-Brick, C. (1991). Lactation, birth spacing and maternal work-loads among two castes in rural Nepal. *Journal of Biosocial Science*, **23**, 137–54.

Panter-Brick, C., Lotstein, D. S. & Ellison, P. T. (1993). Seasonality of reproductive function and weight loss in rural Nepali women. *Human Reproduction*, **8**, 684–90.

Prentice, A. M. (1980). Variations in maternal dietary intake, birthweight and breast-milk output in The Gambia. In *Maternal Nutrition During Pregnancy and Lactation*, ed. H. Aebi and R. G. Whitehead, pp. 167–83. Bern: Hans Huber.

Prentice, A. M. & Prentice, A. (1990). Maternal energy requirements to support lactation. In *Breastfeeding, Nutrition, Infection and Infant Growth in Developed and Emerging Countries*, ed. S. A. Atkinson, L. Å. Hanson & R. K. Chandra, pp. 67–86. ARTS, St. Johns, Newfoundland: Biomedical Publishers.

Prentice, A. M. & Whitehead, R. G. (1987). The energetics of human reproduction. *Symposia of the Zoological Society of London*, **57**, 275–304.

Prentice, A. M., Whitehead, R. G., Roberts, S. B. & Paul, A. A. (1981). Long-term energy balance in child-bearing Gambian women. *American Journal of Clinical Nutrition*, **34**, 2790–9.

Prentice, A. M., Lunn, P. G., Watkinson, M., Lamb, W. H. & Cole, T. J. (1983). Dietary supplementation of lactating Gambian women. II. Effect on maternal health, nutritional status and biochemistry. *Human Nutrition: Clinical Nutrition*, **37C**, 65–74.

Prentice, A. M., Paul, A. A., Prentice, A., Black, A. E., Cole, T. J. & Whitehead, R. G. (1986). Cross-cultural differences in lactational performance. In *Human*

Lactation 2. Maternal and Environmental Factors, ed. M. Hamosh & A. S. Goldman, pp. 13–44. New York: Plenum Press.

Roberts, S. B. & Coward, W. A. (1985). The effects of lactation on the relationship between metabolic rate and ambient temperature in the rat. *Annals of Nutrition and Metabolism*, **29**, 19–22.

Sanborn, C. F., Martin, B. J. & Wagner, W. W. (1982). Is athletic amenorrhea specific to runners? *American Journal of Obstetrics and Gynecology*, **143**, 859–61.

Sanborn, C. F., Albrecht, B. H. & Wagner, W. W. (1987). Athletic amenorrhea: lack of association with body fat. *Medicine and Science of Sports and Exercise*, **19**, 207–12.

Schneider, J., Bradlow, H. L., Strain, G., Levin, J., Anderson, K. & Fishman, J. (1983). Effects of obesity on estradiol metabolism: decreased formation of nonuterotropic metabolites. *Journal of Clinical Endocrinology and Metabolism*, **56**, 973–8.

Schwartz, B., Cumming, D. C., Riordan, E., Selye, M., Yen, S. S. C. & Rebar, R. W. (1981). Exercise-associated amenorrhea: a distinct entity? *American Journal of Obstetrics and Gynecology*, **114**, 662–70.

Shangold, M. M. & Levine, H. S. (1982). The effect of marathon training on menstrual function. *American Journal of Obstetrics and Gynecology*, **143**, 862–9.

Siiteri, P. K. (1981). Extraglandular estrogen formation and serum binding of estradiol: relationship to cancer. *Journal of Endocrinology*, **89**, 119P–29P.

Siiteri, P. K. & MacDonald, P. C. (1973). Role of extraglandular estrogen in human endocrinology. In *Handbook of Physiology*, section 7, vol. 2, part 1, ed. S. R. Geiger, E. B. Astwood & R. O. Greep, pp. 615–29. New York: American Physiology Society.

Singh, J., Prentice, A. M., Diaz, E., Coward, W. A., Ashford, J., Sawyer, M. B. & Whitehead, R. G. (1989). Energy expenditure of Gambian women during peak agricultural activity measured by the doubly-labelled water method. *British Journal of Nutrition*, **62**, 315–29.

Snow, R. C., Barberi, R. L. & Frisch, R. E. (1989). Estrogen 2-hydroxylase oxidation and menstrual function among elite oarswomen. *Journal of Clinical Endocrinology and Metabolism*, **69**, 369–76.

Speroff, L. & Redwine, D. B. (1979). Exercise and menstrual dysfunction. *Physician and Sports Medicine*, **8**, 42–52.

Stallings, J. F., Panter-Brick, C. & Worthman, C. M. (1984). Prolactin levels in nursing Tamang and Kami women: effects of nursing practices on lactational amenorrhea. *American Journal of Physiological Anthropology*, Supplement 18, 185–6.

Stein, Z. & Susser, M. (1978). Famine and fertility. In *Nutrition and Human Reproduction*, ed. W. H. Mosely, pp. 123–46. New York: Plenum Press.

Stern, J. M., Konner, M., Herman, T. N. & Reichlin, S. (1986). Nursing behaviour, prolactin and postpartum amenorrhea during prolonged lactation in American and !Kung mothers. *Clinical Endocrinology*, **25**, 247–58.

Tremblay, A., Despres, J-P., Leblanc, C., Craig, C. L., Ferris, B., Stephens, T. & Bouchard, C. (1990). Effect of intensity of physical activity on body fatness and fat distribution. *American Journal of Clinical Nutrition*, **51**, 153–7.

Van Raaij, J. M. A., Schonk, C. M., Vermaat-Miedema, S. H., Peek, M. E. M. & Hautvast, J. G. A. J. (1991). Energy cost of lactation, and energy balances of well-nourished Dutch lactating women: reappraisal of the extra energy requirements of lactation. *American Journal of Clinical Nutrition*, **53**, 612–19.

Veldhuis, J. D., Evans, W. S., Demers, L. M., Thorner, M. O., Wakat, D. & Rogol, A. D. (1985). Altered neuroendocrine regulation of gonadotropin secretion in women distance runners. *Journal of Clinical Endocrinology and Metabolism*, **60**, 557–63.

Wakat, D. K., Sweeney, K. A. & Rogol, A. D. (1978). Reproductive system function in women cross country runners. *Medicine and Science of Sports and Exercise*, **14**, 263–9.

Walker, S. M., Walker, R. F. & Riad-Fahmy, D. (1984). Longitudinal studies of luteal function by salivary progesterone determinations. *Hormone Research*, **20**, 231–40.

Warren, M. (1980). Effects of exercise on pubertal progression and reproductive function in girls. *Journal of Clinical Endocrinology and Metabolism*, **51**, 1150–7.

WHO (1985). The quantity and quality of breast-milk. *Report of the WHO Collaborative Study on Breast Feeding*. Geneva: World Health Organisation.

Worthman, C. M., Jenkins, C. L., Stallings, J. F. & Lai, D. (1993). Attentuation of nursing-related ovarian suppression and high fertility in well-nourished, intensively breastfeeding Amele women of lowland Papua New Guinea. *Journal of Biosocial Science*, **25**, 425–43.

Index

Lightning Source UK Ltd.
Milton Keynes UK
UKOW041806170413

209389UK00001B/43/P

05791858